AF713130

WHO was WHO in COMPANY K

Reliable facts about the Native American soldiers in Company K, 1st Michigan Sharpshooters, during the Civil War 1861-1865

Researched by Chris Czopek
~ Lansing, Michigan ~

Second Edition (P3) © 2015

Red Oak Research: http://RedOak1863.homestead.com

Copyright © 2015 by Chris Czopek

Published by **Red Oak Research**

ISBN: 978-0-9969474-0-4
EAN 13: 0-9969474-0-X

All rights reserved. No part of this book may be reproduced in any form or by any electronic or mechanical means, including information storage & retrieval systems, scanning, or electronic distribution without permission *in writing* from the author, except by reviewers who my quote brief passages in a review.

For information, contact Chris Czopek using this email: redoak1863@aol.com

DEDICATION –

The State Capitol building in Lansing, Michigan, is visited each year by thousands of schoolchildren. Volunteers give tours of the inside, and if the weather is good, the students walk outdoors to see the Civil War monument on the front lawn. Standing before the stone soldier, they hear the story of the all-Indian unit, called "Company K". Being children, they have questions about those Michigan Indians who fought in the Civil War – What did their names sound like? What tribes were they from? How many died in battle?
To those curious children, I dedicate this book.

Picture: "The Sharpshooter" on front lawn of State Capitol. Photo by Chris Czopek – Lansing, MI

AKNOWLEGEMENTS

I owe a debt of gratitude to author and historian Raymond Herek. When his book "These Men Have Seen Hard Service" was published in 1998 by Wayne University Press, it immediately became the best source of information about Company K and the 1st Michigan Sharpshooters regiment. His book is one of my main sources of information, and I acknowledge my reliance on his groundbreaking research.

Another author and historian deserving mention is James N. Jackson. Jim is a quiet, patient scholar who has made a number of important discoveries. For example, he discovered the truth behind the arrest of Sgt. Charles Wabesis in Chicago. I consider Jim Jackson one of the leading experts on Company K, and his willingness to trade information is gratefully acknowledged.

The late Dr. Charles J. Meyers was the real pioneer of Company K research. As an Odawa boy growing up in the Petoskey area, he listened to the grandsons of the soldiers tell war stories that were never written down. This gave him a unique perspective on the story of Company K. During two long phone calls one summer, he gave me valuable information and insight. It is unfortunate Charles passed away before finishing his book.

Assisting Dr. Meyers was a Lansing historian named Roger Winthrop. He continued the work Charles started, and became something of an expert in his own right. Roger was the first to discover the true fate of poor Daniel Ashman. For this and other information he shared with me, I am grateful.

Don Otto, grandson of Marcus Otto, and the Anishinabe Ogitchedaw Veterans generously supported my work to honor the Company K soldiers. Chi Meegwetch !

Thanks to the others who helped me over the years, including Michigan's leading expert on Native American genealogy, Barbara Madison. Nobody knows the family lines of Company K soldiers better, and her corrections and facts were invaluable. She took time from her busy schedule to help me, and I will always be grateful. My thanks to another expert on Native American genealogy, Jim LaLone. His focus is on Michigan's French-Canadian families. In his files I found many of the soldiers of Company K, along with facts on family connections not available anywhere else.

Deserving special thanks is the late Avis Wolfe of Northport, guardian of the Wolfe family papers and stories. Her transcription of the diary of missionary George Nelson Smith was a wonderful source of information on the Company K men who came from the Leelanau peninsula. She donated her transcription of the diary, and other valuable items inherited from Payson Wolf's family, to the excellent little museum in Northport, Michigan. To anyone doing serious research on Company K, I recommend a visit to this museum.

Constance Larson generously shared her remarkable collection of information. I also thank GTB members Art Dembinski M.D., an expert on the Wakazoo family, and Scott Schwander, who tirelessly searches the internet for new information. Victoria Wilson, genealogist, created the first website on the internet about Co. K. She is still active today, leading an organized search for new information.

Jack Soper of Bath, Michigan, was the first to find Joe Hannin in Canadian records, and establish Joe's Iroquois - Oneida background (a major discovery!)

Paul M. Wandrie Sr. of Mackinac Island deserves credit for discovering the post-war enlistment of Benjamin Kewaconda in the army.

John Hughes of Battle Creek gave me my first lesson in researching Company K soldiers.

Bill Skillman began as an expert on the Berdan Sharpshooters, then took an interest in the First Michigan Sharpshooters; he discovered many nuggets of information. Fond memories of our annual meetings in Jackson before a campfire.

Quita V. Shier of Midland, Michigan, spent 26 years collecting information on the soldiers of Company K. In December of 2017, she printed all her notes in a 550 page self-published book titled "Warriors in Mr. Lincoln's Army". She obviously borrowed information from an early edition of my book "Who Was Who In Company K" without mentioning the source. In return, I have borrowed a few facts from her book. But each time I did, credit is clearly given to her as the "first" to discover and publish the information.

My biggest thanks go to my patient wife, Bonnie. On many research trips she rode with me in the car, reading aloud from history books. It was her constant love, understanding and support that made writing this book possible.

Who was Who in Company K

TABLE OF CONTENTS

Introduction to COMPANY K – the Forgotten Story

During the Civil War, the 1st Michigan Sharpshooters regiment was being recruited to fight for the Union. But there was a problem – few men could pass the marksmanship test. Since Native Americans were famous as skilled hunters, it was decided to recruit one company - Company K - from among the tribes in Michigan. Nearly 140 men volunteered for Company K. They had names like Wolf, Greensky, Kakakee (Sparrow Hawk), Agahgo (Old Porcupine), and Wakazoo.
They were sent to Virginia in 1864, where they fought in some of the fiercest battles of the Civil War: The Wilderness, Spotsylvania, and the terrible Siege of Petersburg. During an attack against the enemy lines June 17, 1864, a group of 15 men were captured and sent to the infamous Andersonville Prison. The rest of the company fought on till the end of the war. They proudly marched in the famous Grand Review, the official victory parade in Washington D.C. Afterwards they were mustered out of the army, sent home to Michigan, and … forgotten.

The inspiration to write this book came from a phone call. A student at a Tribal College was taking a history class. For a final project, the teacher passed out a list of possible subjects. This student chose Company K.

On the phone, she told me how difficult it was to find any information at the local libraries. Even a university like Michigan State, which offers a Native American studies program, has nothing in their library on the subject of Company K.
So she called me on the phone to ask for help. I sent her a thick manila envelope full of notes. Once she had the information, she was able to write an excellent research paper. From this phone call, I learned there are new Tribal Colleges being created on reservations, and state universities are expanding their Native American history programs. The student I helped was part of a new generation taking history classes taught by Native American teachers. These students will someday become historians and scholars. They will rewrite the history of Michigan's Native Americans.
If Company K is going to be part of that history, these students will need a good source of information. That call convinced me, it was time to publish this book.

There has long been a need for solid, accurate information on the soldiers of Company K. No two sources agree on how many men enlisted, how their names were spelled, or the fate of each man in the war. And whenever I tell the story to a history group, they always ask: What happened to these men after the war?
This is the first book to answer these questions. It is a "guide book" for students and historians exploring the history of Company K.

Chris Czopek
Lansing, Michigan

My Sources Of Information

Much of the information in this book came from the Michigan Library and Archives in Lansing. I know the place well. For over 25 years I have searched through its books, microfilm collections, and neatly labeled boxes of papers. To better explain the footnotes in this book, please read the following list, my Ten Best Sources for information on Company K soldiers.

THE ADJUTANT GENERAL RECORD

In 1907, all the service records of all the Michigan soldiers in the Civil War were published in a 46 volume set of books called Record of Service of Michigan Volunteers in the Civil War 1861-1865. Since the information came from records in the office of the state Adjutant General, historians and genealogists call these books the "**Adjutant General Record**".
All the service records for the Company K men are found in Volume 44.
The official army record for every soldier in Company K is his Adjutant General Record. Here is a sample:

Pe-nais-now-o-quot, Jacko, Northport. Enlisted in company K, First Sharpshooters, June 1, 1863, at Northport, for 3 years, age 40. Mustered June 22, 1863. Taken prisoner at Petersburg, Va., June 14, 1864. Died in prison at Andersonville, Ga.

THE ADJUTANT GENERAL ANNUAL REPORT

Every state has a National Guard. The man in charge of it is called the State Adjutant General, and his office keeps the records of the soldiers.
Once a year, the Adjutant General for the state of Michigan publishes an annual report. During the Civil War years, these reports were large, thick books, and they contained long lists of names of Michigan soldiers. This annual report gives valuable information about the men, such the home county of the soldier, and his fate (Killed, Captured, Discharged, etc.).
While the Adjutant General Records also has this information, some of the facts on the Annual Report are different, as if copied from different sources. This makes it valuable for double-checking information and spotting errors.
Here is a sample:

Died of Disease, &c.

NAMES.	CO.	COUNTY WHERE HE RESIDED.	REMARKS.
Gibson, Joseph L..........	K	Canada,............	In Mil. Prison, at Andersonville, Ga.
Kadah, Joseph.............	"	Oceana,............	At St. Mary's Hosp., Detroit, Oct. 23, 1864.

PENSION FILES – the best source of information for Company K

A Pension File is a file folder with the soldier's name on it. Inside are many different kinds of documents. Some are hospital records telling what day the soldier was wounded, and where (what battle) and exactly what the injury was. Other documents are from widows asking for pensions, giving date of marriage, place of marriage, and a list of children. Native American soldiers were sometimes married by "native custom" so there is no written record. In these cases, to give the government documented proof of marriage, two witnesses to the marriage would go to the county clerk's office, and give sworn testimony on when and where the marriage took place. The document would be sent to Washington D.C. and placed in the Pension File. Since children under the age of 16 could receive pensions, the government would ask for a list of children and their birthdays. That too would go to the pension file. For Native American genealogy, a Pension File is a gold mine of family information that cannot be found anywhere else!

See Appendix A of this book for a complete list of all the pension files for the Company K men. Pension Files are kept at **National Archives** in Washington D.C. You can order copies from their website on the internet.

Here are the rules for Civil War Pensions:

- **INVALID PENSION:** If a soldier was wounded or injured in the war, he could apply for an Invalid Soldier Pension.
- **INVALID PENSION for old soldiers:** After 1900, any soldier still alive and suffering from old age could receive an Invalid Soldier Pension.
- **WIDOW PENSION:** If a soldier was killed in the war, his widow could apply for a Widow Pension. If she remarried, her pension stopped on her wedding day.
- **MINOR PENSION:** If a soldier was killed in the war, any of his children under the age of 16 could receive a Minor Pension. If their mother remarried and lost her pension, her children still collected their pensions.
- **MOTHER / FATHER PENSIONS:** It happened several times, that the mother of a soldier in Company K who died in the war received a pension. She had to prove her son was her sole means of support, and she was unable to work and earn her own living. Fathers could also apply for a pension, for the same reasons.
- **NOTE on SOLDIERS WHO DIED AFTER THE END OF THE WAR:**
  If a veteran of Company K was discharged from the army and died soon afterwards, his widow and children could collect pensions if they could prove his death was directly caused by sickness or injury received during the war.
- **NOTE on APPLICATION NUMBER and CERTIFICATE NUMBER :**
  In Appendix A of this book, you will see that every pension has two numbers, because there were two steps to getting a pension. Step 1, a soldier or widow sent an application to Washington DC and they started a file with an **Application number**. Step 2, when the pension is approved, they issue a Certificate, and the number on the pension file is changed to a **Certificate number**. If the Certificate number is blank, the pension was not granted.

1890 VETERAN CENSUS

When the U.S. Census was being taken in the summer of 1890, the Census Taker would ask, “Is there a Civil War veteran living here?” If the answer was yes, he would take out a separate sheet of paper and ask for the soldier’s Regiment, Enlistment Date, Discharge Date, and wounds/injuries. If a widow was living in the home, he would ask these same questions, and make a note of the name of the widow. For tracking down Company K soldiers after the war, this census is a valuable source of information!

The 1890 Veteran Census is on microfilm, and is available at libraries in Michigan’s largest cities. Look for it in their genealogy collection under the name “1890 Federal Census of Soldiers, Sailors and Widows”

My book, **Who Was Who in Company K**, will tell you which soldiers appear on the 1890 Veteran Census, and how they spelled their names. See Appendix B.

Warning: Some people will tell you the 1890 census does not exist, that it was destroyed. That is true for the regular US census, but this is different, this is the Veteran Census of 1890.

1888 VETERAN CENSUS

This is a rare, little known source of information on Civil War soldiers.
It just had three bits of information: Name, Regiment & Company, and Post Office. For researching Company K soldiers, this will tell you who was alive in 1888, how they spelled their name, and where they were living.

It is available only at the Archives reading room of the State Library and Archives in Lansing. Ask for Reel # 684, “Names and Addresses of Soldiers Living In The State of Michigan in 1888”. My book, **Who Was Who in Company K**, will tell you which soldiers appear on the 1888 Veteran Census. See Appendix B.

1894 VETERAN CENSUS

The state of Michigan used to do a state census every ten years. In 1894 they asked at each house if there was a veteran living there. This source just has just two bits of information: Name and town/township.

For researching Company K soldiers, this will tell you who was still alive in 1894, how they spelled their names, and where they were living.

It is available as a hardcover book at the Library of the State of Michigan in Lansing. It also can be found at libraries in many of Michigan’s larger cities.

My book, **Who Was Who in Company K**, will tell you which soldiers appear on the 1894 Veteran Census list, and how they spelled their names. See Appendix B.

DESCRIPTIVE ROLL, State Archives

The original record books kept in the Adjutant General's Office were called "The Descriptive Roll". Every Michigan soldier in the Civil War has his service record handwritten in this book. Most historians ignore this source, thinking all the information was published in the 46 volumes set of books called Record of Service Of Michigan Volunteers in the Civil War 1861-1865. But those historians are mistaken!

The Descriptive Roll gives additional information such as Occupation, Residence, and the County to be credited for draft purposes (a useful way to find where their home really was, and what tribe they were from).

The Descriptive Roll also has notes on sickness, temporary duty assignments such as hospital orderly, and the promotions / demotions of the soldier.

No study of the soldiers in Company K is complete without looking at the Descriptive Roll. You can find it only at the state archives in Lansing. Ask in the reading room for Microfilm Reel No. 593.

GENDIS - DEATH RECORDS FROM 1867 to 1897

Starting 1867 the Department of Health for the state of Michigan required all deaths be reported to Lansing. Many veterans from Company K show up on these reports. This is valuable information, because when you know the date and place of death, you can search for an obituary in local newspapers.

These Death Records are on microfilm at state archives in Lansing, but they are also on line. Look on the internet for GENDIS (= GENEALOGY DEATH INDEX SYSTEM) www.mdch.state.mi.us/gendisx/index.htm

STATE DEATH RECORDS (1897-1920)

These Death Records are on line at www.seekingmichigan.gov

Besides giving the date and place and cause of death, these records tell what cemetery the person is buried in. Also, they may have the name of the parents, making them an excellent source of information for Company K soldiers.

Sample of part of a State Death Record for Company K soldier Joe Kakakee:

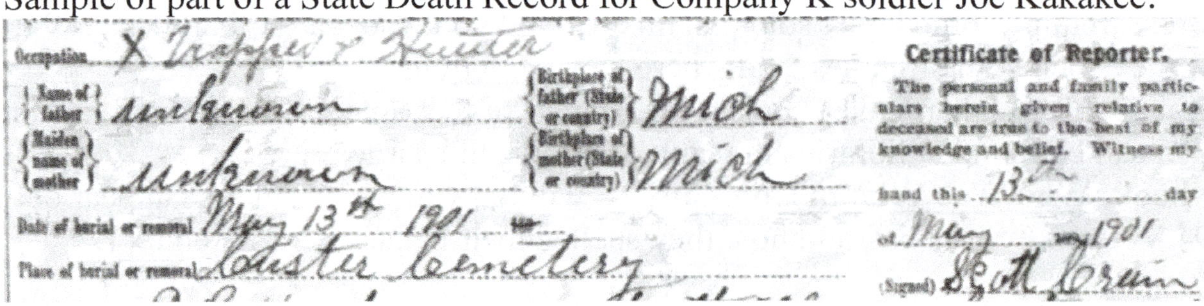
Occupation X Trapper & Hunter
Name of father unknown — Birthplace of father (State or country) Mich
Maiden name of mother unknown — Birthplace of mother (State or country) Mich
Date of burial or removal May 13th 1901
Place of burial or removal Custer Cemetery

Certificate of Reporter.
The personal and family particulars herein given relative to deceased are true to the best of my knowledge and belief. Witness my hand this 13th day of May 1901
(Signed) Scott Crain

Occupation: Trapper & Hunter. Buried May 13, 1901 in Custer Cemetery.

1883 PENSION ROLL

In 1883, the government published a list of all the soldiers & widows receiving pensions. Its official name is List of Pensioners on the Roll January 1, 1883. It was printed as a five volume set of books. Volume 4 has a list of ALL the people in Michigan receiving pensions from the government. There are 19 soldiers & widows from Company K on this list.

For each of these 19 people, it shows the County & Post Office where they were living, their pension certificate number, date the pension was granted, and the reason they receives a pension. (Example: g.s.w. is an abbreviation for gun shot wound.)

A copy of this book is at State of Michigan Library: E 494 / .U55 1883.

My book, **Who Was Who in Company K**, will tell you which soldiers appear on the 1883 Pension Roll. See Appendix B.

Book:

THESE MEN HAVE SEEN HARD SERIVICE; The First Michigan Sharpshooters in the Civil War, by Raymond J. Herek.

Wayne State University Press, Detroit, 1998.

The author, Ray Herek, wrote this book about the 1st Michigan Sharpshooters regiment. Since Company K was a part of this regiment, there is much information about the Native Americans. A handy index in the back of the book will let you look up specific names. Between the covers of this book is the story of Company K from beginning to end, told by a first rate historian.

Book:

WARRIORS IN MR. LINCOLN'S ARMY; Native American Soldiers Who Fought In The Civil War, by Quita V. Shier.

iUniverse, Bloomington, Indiana, December 2017.

The author, Quita V. Shier, spent 26 years gathering genealogy records on the men of Company K. She also examined their pension files and medical records. The result is this book – a database focused on family history and medical history.

Magazine Story:

INURED TO HARDSHIPS, FLEET AS DEER, by Gordon Berg.

Published in Civil War Times, June 2007

Copy available at the Ziibiwing Cultural Center, Mt. Pleasant, MI.

Magazine Story:

THE BRAVERY OF COMPANY K, by Elizabeth Edwards.

Published in Traverse Magazine, February 2003

Copy available at the Ziibiwing Cultural Center, Mt. Pleasant, MI.

Frequently Asked Questions

Q: What is COMPANY K ?

Company K was an all-Indian company of Civil War sharpshooters.
It has a special history: In 1863, a regiment of Sharpshooters was being recruited in Michigan to fight for the Union, but there was a problem – few men could pass the marksmanship test. Since Native Americans were famous as skilled hunters, it was decided to recruit one company from among the tribes in Michigan. A total of 139 men volunteered to join.

Each man passed the test, hitting a 5 inch circle from a distance of 220 yards. For Basic Training, they were sent to Dearborn, Michigan.
The soldiers of Company K wore the same uniform and received the same pay as the rest of the regiment. The newspapers in Detroit watched them go through Basic Training and reported they made excellent soldiers. In 1864 Company K was sent to Virginia, where it fought under General Grant in some of the fiercest battles of the Civil War.

After Lee surrendered, the surviving soldiers of Company K marched in the famous Grand Review, the official victory parade in Washington D.C.

The story of Company K was forgotten till publication in 1998 of These Men Have Seen Hard Service, a book by historian Raymond Herek.

Company K was part of the 1st Michigan Sharpshooters, a Michigan infantry regiment. A monument to these Sharpshooters is on the front lawn of the state Capitol building in Lansing.

Q: Was every soldier in Company K a Native American ?

A: Yes. Every enlisted man in Company K was a Native American, including Henry G. Graveraet Jr., the First Sergeant.
All the officers in Company K were white except for Lt. Garrett Graveraet, who was an Odawa (Ottawa) from Mackinac Island.

Q: How many men were in Company K ?

A: 145 soldiers served in Company K during the Civil War.
When you subtract the 6 white officers, you have 139 Native Americans.

Q: What tribes did they come from ?

A: 53% Odawa (Ottawa), 40% Ojibwa (Chippewa), 5% Bodewadimi (Potawatomi), 2% unknown.

Q: Historians often ask, "How many soldiers have Pension Files?"

A: Of the 139 Native American soldiers, 102 have pension files.

Q: What battles did the Company K soldiers fight in ?
A: Company K fought in Virginia 1864-1865 as part of 9th Army Corps. Biggest battles: The Wilderness, Spotsylvania, and the Battle of the Crater. They took heavy casualties June 17, 1864 during the Siege of Petersburg. (No, they did not fight at Gettysburg or March to the Sea with Sherman.)

Q: How many men in Company K died in the war ?
A: Out of 145 men, 46 died during the war (32 % of the Company). There were 12 soldiers Killed In Action, 9 died of battle wounds, 16 died from disease or accidents, and 9 P.O.W.s who died of starvation & disease.

Q: How many Purple Hearts ?
A: By today's standards, 63 soldiers earned the Purple Heart. (That's 45% of the Company).

Q: Were there any Company K soldiers at Andersonville prison ?
A: Yes. 15 men were captured June 17, 1864 and sent to Camp Sumter at Andersonville, GA. Half these men died of starvation and disease.

Q: Were there any Company K soldiers on the steamboat Sultana ?
A: Yes, two men were on the SULTANA when it exploded and burned on the night of April 27, 1865. Their names were Amos Ashkebugnekay and Louis Miskoguon. Both men survived the tragedy.

Q: Were there any heroes in Company K ?
A: Yes. Major General John G. Parke, commander of 9th Army Corps, personally recommended Antoine Scott for the Medal of Honor.

Q: What kind of rifles did these Sharpshooters use ?
A: The soldiers of Company K used the standard Springfield rifle, with no special gunsight. During idle time, individual soldiers carved into the stocks pictures of fish, snakes, and animals.

Q: Who was the last surviving member of Company K ?
A: Moses Thomas was the last surviving member of Company K. He died on the reservation at Kettle Point, Canada, in 1933.

Q: Has anyone ever made a movie about Company K ?
A: Yes, an award-winning documentary titled "Road To Andersonville" by David B. Schock Ph.D. First premiered in 2013, it tells the entire story of Company K from start to finish. Highly recommended.

Q: Was any soldier in Company K an " Indian Chief " ?
A: Yes, one - Daniel Mwakewenah. He signed the Treaty of Saginaw. This 40 year old chief urged the men of his tribe, the Bear River band of the Odawa (Ottawa) tribe, to volunteer for Company K. And then he went with them to share their fate. He died from wounds received in battle.

Q: Were all the soldiers in Company K from Michigan ?
A: 98% were residents of Michigan at the time they enlisted in Company K. There were two from Ashland County, Wisconsin, who crossed the state line to enlist in Michigan. And two or three men from Canada traveled to Michigan to enlist, and after the war returned to Canada. One volunteer, Charles Carter, was from Hagerstown, Maryland (Delaware tribe?).

Q: Are there any Company K soldiers buried in Arlington ?
A: According to official records, three men from Company K are buried in Arlington National Cemetery near Washington D.C. They are:
Pvt. Oliver Aptargeshick in Section 13, Grave 5732.
Sgt. Peter Burns in Section 13, Grave 8141.
Pvt. David George in Section 27, Grave 325.
All three were wounded in battle and died during the war.

Q: Did any of the veterans of Company K join the G.A.R. ?
A: The G.A.R. was a Civil War veterans group formed after the end of the war. It was like the V.F.W. or American Legion of today.
Yes, 19 veterans of Company K joined their local G.A.R. posts.
Some posed for photographs proudly wearing their G.A.R. badge.
Note: The Mt. Pleasant WA-BU-NO G.A.R. Post No. 250 was named for Thomas Wabuno, a Company K soldier who died during the war.

Q: How many Native Americans from Michigan volunteered to fight in the Civil War ?
A: About 175. Here is how I come up with this number:
There were 139 in the 1st Michigan Sharpshooters Co. K, and two in Company B.
In the 9th Michigan Cavalry Co. K, there were several.
A few Ojibwe (Chippewa) from the Swan Creek & Black River bands joined the 10th Michigan Infantry Co. B.
And I have run across more than twenty other Native Americans who joined various regiments. By my count, the total is around 175.

Q: In the army, what exactly is a Company ? What is a Regiment ?
A: A company is a unit of 100 men commanded by a Captain and two Lieutenants.
A regiment is a unit of 1000 men divided into ten companies and commanded by a Colonel.
The Companies in a regiment are named A,B,C.D,E,F,G,H,I,K. (Notice, there is no Company J). So, Company K was one of ten companies in a regiment called the 1st Michigan Sharpshooters.
Note: After a few battles, no company ever had 100 men.

Q: What are the Army Ranks ?
A: The ranks of soldiers are divided into two groups: Officers and Enlisted.
OFFICERS
General (Four Stars). There were no 4 star generals in the Civil War.
General (Lieutenant General, Three Stars) – Commander of Union Army.
General (Major General, Two Stars) - Commander of a Division or Corps.
General (Brigadier General, One Star) - Commander of a Brigade.
Colonel - Commander of a Regiment, like the 1st Michigan Sharpshooters.
Lieutenant Colonel - takes over when the Colonel is away.
Major - takes over when the Lt. Colonel is away.
Captain - Commander of a Company.
First Lieutenant - Helps the Captain run the Company.
Second Lieutenant - Lowest ranking officer in the army.

ENLISTED MEN
1st Sergeant - in charge of all the Sergeants in the Company.
Sergeant - think of him as a strict supervisor.
Corporal – think of him as an assistant supervisor.
Private - The common soldier. Lowest rank in the army.

Q: Who was the Commander of Company K ?
A: First, Captain Edwin Andress was in command. He was a white man from Chesaning who knew the language and customs of Native Americans. A bullet at Spotsylvania ended his army career. Then Captain James DeLand took over as commander of Company K. He was the brother of Colonel DeLand. The men of Company K felt honored that the Colonel chose his own brother to lead them into battle.

Q: Some old records say the majority of the men in Company K were recruited in the Upper Peninsula. Is that true?
A: False! Less than 10 men enlisted in the Upper Peninsula.

Q: Are there any monuments to the soldiers of Company K ?
On the front lawn of the Capitol building in Lansing, there is a monument to the 1st Michigan Sharpshooters regiment - on a bronze plaque attached to this monument is the name of Lt. Garrett Graveraet.
There is a memorial stone for Oliver Aptargeshick on Walpole Island, Canada. The name Sashkobonquot was added to the Veterans Memorial in front of the Bay County Building in Bay City on May 30, 2011.
A memorial to Civil War soldiers in Traverse City on 11th street and Elmwood has the names of five men from Company K.
Lake City in Missaukee County has the name Cornelius Hall on the new Veterans monument (on the corner of Main St. and Prospect St).
A small bronze plaque on the wall inside the Main Gate of the historic Dearborn Arsenal (just outside the old Commandant's Quarters on the corner of Monroe street and Michigan street) mentions Company K.
There is a Veterans Memorial in Hart, Michigan (Oceana County) with the names of 12 local men from Company K carved in stone.
Beaver Island has a Veterans Memorial dedicated August 1, 2015, with bricks for James Washegesic and Joseph Shawonosang. Mackinac Island dedicated a monument in 2017 to its Civil War soldiers, two Co. K men.
Petersburg, VA: Plaque on Peter Jones Trading Station building mentions Co. K. prisoners held there; located on corner of North Market & Grove.

Q: Enlistment records show that many men came from Bear Creek and Little Traverse. Where are these places on a map ?
A: Bear Creek changed to Petoskey, Little Traverse to Harbor Springs.

Q: Did any soldiers transfer into Company K?
A: Just one - Joseph Wakazoo. He first joined the 16th Michigan Infantry.

Q: Did any soldiers transfer out of Company K ?
A: Thomas Nelson was injured, so he transferred to the V.R.C. (Veteran Reserve Corps) to do light duty such as security guard at an army hospital.

Q: Collectors ask, are there any Civil War uniforms or rifles passed down in the family of a Company K soldier ?
A: No. However, when John Rubberdee was discharged, he paid $6.00 so he could keep his rifle and take it home. Where is that rifle today?

Q: Who were the Lieutenants in Company K and why do they matter ?
A: Besides Lt. Garrett Graveraet, there was: William J. Driggs, Leumuel R. Nichols, Kirk W. Noyes, and Winfield S. Shanahan. Their descendants may have letters written during the war that talk about the Company K soldiers.

ALL NAMES INDEX

This index will guide you through the many, many spellings of the names of the soldiers.
It is intended for those students and scholars willing to search old records for <u>new information.</u>

There is no correct spelling for these names. In bold type is the "prime name". It is the most commonly used name for the soldier. The names in regular type are all "aka" names followed by a number.

The CL number is a sort of army serial number for each soldier. This number will lead you to my notes about his life and death.

<u>ALL NAMES INDEX OF COMPANY K SOLDIERS</u>

 All Rights Reserved by Chris Czopek

ALL NAMES INDEX OF COMPANY K SOLDIERS

Battice, John = CL-014
Benasis, John = CL-015
Bennett, Louis = CL-016
Bensus, John = Benasis, John CL-015
Beuchamp, Augustus = Boushaw, Augustus CL-017
Bird, John = Benasis, John CL-015
Boo Shaw, Charles = Shaw, Charles CL-113
Boushaw, Augustus = CL-017
Burns, Peter = CL-018
Bushaw, Augustus = Boushaw, Augustus CL-017
Cabecoung, William = CL-019
Carter, Charles = CL-020
Cay-bar-coung, William = Cabecoung CL-019
Cay-bar-cowing, William = Cabecoung, William CL-019
Chabin = Cabecoung, William CL-019
Chalfield, Samuel C = Chatfield, Samuel C CL-023
Chamberlain, Amos = CL-021
Chamberlin, Amos = Chamberlain, Amos CL-021
Chatfield, Charles = CL-022
Chatfield, Samuel C = CL-023
Chebbin = Cabecoung, William CL-019
Chetego, Thomas K = Kechittigo, Thomas CL-063
Chibadice, A. = Ketchebatis, Amable CL-066
Chippen = Cabecoung, William CL-019
Chirch, Albert = Church, Albert CL-024
Chittago, Thos = Kechittigo, Thomas CL-063
Church, Albert = CL-024
Church, James = Church, Albert CL-024
Collins, Jacob = CL-025
Collins, John = CL-026
Collins, William = CL-027
Condecon, Henry = Wassagezic, Henry CL-133
Corbaicong, William = Cabecoung, William CL-019
Corbin, George = CL-028
Cowing Cayba, William = Cabecoung, William CL-019
Cowing, William C = Cabecoung, William CL-019
Crane, Amos = CL-029
Crittago, Benj. = Kechittigo, Thomas CL-063
Dabasequam, Jonah = CL-030
David, John = CL-031
Davis, Jno = David, John CL-031
Deverney, John = Kedgnal, John CL-146
Dutton, Luke = CL-034
Dutton, Luther = Dutton, Luke CL-034
Duvernay, John Henry = Kedgnal, John CL-146
Duvernay, William = CL-147
Duverney, "Chid" = Duvernay, William CL-147
Egawanige, Amabilis = Ketchebatis, Amable CL-066
E-tar-ivegezhik, John = Etarwegeshig, John CL-035
Etarwegeshig, John = CL-035
Gagwich, Samuel = Kaquatch, Samuel CL-062
Gaveraet, Garrett A. = Graveraet, Garrett A. CL-040
Genereau Jr, Louis = CL-036
Genereau Jun., Louis = Genereau Jr, Louis CL-036
Genereaux, Louis = Genereau Jr, Louis CL-036
George, David = CL-037
Gibson, J. L. = Gibson, Joseph L. CL-038
Gibson, Joseph L. = CL-038
Gibson, Louis G. = Gibson, Joseph L. CL-038
Gijigowi, Simon = Kejikowe, Simon CL-065
Gisecomey, William = Waygeshegoing CL-137
Goeng, Samuel = Going, Samuel CL-039
Going, Samuel = CL-039
Gora, Samuel = Going, Samuel CL-039
Gorgo, Charles A = Agahgo, Charles CL-001
Graveraet Jr, Henry G. = CL-041
Graveraet, Garrett A. = CL-040
Graverates, Henry = Graveraet Jr, Henry G. CL-041
Graverel, H. G. = Graveraet Jr, Henry G. CL-041
Green, Amos = Ashkebugnekay, Amos CL-011
Green, Philip = Ashkebugnekay, Amos CL-011
Greensky, Benjamin C. = CL-042
Greensky, Jacob = CL-043
Greuitt, Peter = Gruet, Peter CL-044
Grewett, Peter Jr. = Gruet, Peter CL-044
Grewitt, Peter = Gruet, Peter CL-044
Groit, Peter = Gruet, Peter CL-044
Grovemet, Henry = Graveraet Jr, Henry G. CL-041
Gruet, Peter = CL-044
Haegler, Joseph = Kadah, Joseph CL-060
Hagler, Joseph = Kadah, Joseph CL-060
Hall, Cornelius = CL-045
Hamblin, James M. = Hamlin, James H CL-046
Hamlin, J H = Hamlin, James H CL-046
Hamlin, James H = CL-046
Hannin, Joseph = CL-047
Harnoist, Joseph = Hannin, Joseph CL-047
Heighton, Joseph = Kadah, Joseph CL-060
Heogler, Joseph = Kadah, Joseph CL-060
Herbert, Charles = Hubert, Charles CL-048
Honyust, Joseph = Hannin, Joseph CL-047
Hornicost, Joseph = Hannin, Joseph CL-047
Hubert, Charles = CL-048
Ioco, John = Jacko, John CL-051
Isaac, Wm = Isaacs, William CL-050
Isaacs, John = CL-049
Isaacs, William = CL-050

 All Rights Reserved by Chris Czopek

ALL NAMES INDEX OF COMPANY K SOLDIERS

 All Rights Reserved by Chris Czopek

Mwakewenah, Daniel = CL-082
Mwa-ke-wenah, Daniel = Mwakewenah CL-082
Nabawnayasang, Joseph = Gibson, Joseph CL-038
Na-baw-na-ya-sang, Joseph = Gibson CL-038
Nab-en-ay-ash, T. C. = Smith, Thomas CL-120
Nai-tak-me-no-ting, Jacko = Jacko, Nat. CL-052
Naquam, Thompson = Nauquam, Thompson CL-085
Narquaquot, Joseph = CL-083
Nar-ue-ge-she-gwa-by, Jackson = CL-084
Narwegeshequabey, Jackson = CL-084
Nar-we-ge-she-quaby, Jackson = CL-084
Natawinoding, Louis = Jacko, Nat. CL-052
Nat-tah-me-no-ting Jacko = Jacko, Nat. CL-052
Nauquam, Thompson = CL-085
Naw-yah-waw-sung = Wesley, John CL-140
Nelson, Thomas = CL-086
Nesogot, Joseph = CL-087
Nevarre, William = Neveaux, William CL-088
Neveaux, William = CL-088
Newton, William = CL-089
Now-quo-um, Charles = Whiteface, Charles CL-141
Odesheweiad, Leo = Otashquabono, Leon CL-093
Ohbowakemo, James = CL-092
Olwayquonske, James = Ohbowakemo CL-092
Oquarboway, James = Quoboway, James CL-103
Osah-ogash-you = Sawbequom, Adam CL-108
Osah-o-quah-yon = Sawbequom, Adam CL-108
Oshburn, James R. = Ohbowakemo CL-092
Otashquabono, Leon = CL-093
Otosgabench, Leone = Otashquabono, Leon CL-093
Otto, Marcus = CL-094
Otto, Solomon = CL-095
Pa sau o quot, Jacko = Penaiswanquot, Jacko CL-098
Pakemaboga, John = CL-096
Parceley, Charles = Porcellay, Charles CL-101
Pashabe, John = Kinewahwanipi, John CL-068
Pasharba, Albert = Pesherbay, Albert CL-100
Pay-mos-se-gay, Daniel = Pemassegay CL-097
Pay-wau-su-quot = South, Peter CL-121
Pe-che-ka, Louis = Peshekee, Mark CL-099
Pemassegay, Daniel = CL-097
Penaiswanquot, Jacko = CL-098
Penaiswanquot, Louis = Jacko, Nat. CL-052
Penaiswonquot, Jake = Penaiswanquot CL-098
Penaswonquot, Jaco = Penaiswanquot CL-098
Peneaswonoquot, Jocko = Penaiswanquot CL-098
Penegwah, John = Kinewahwanipi, John CL-068
Peqongay, Aaron = Sahgahnahquato CL-105
Pequongay, Aron = Sahgahnahquato CL-105
Pequongway, Aaron = Sahgahnahquato CL-105

ALL NAMES INDEX OF COMPANY K SOLDIERS

Peshawbey, Albert = Pesherbay, Albert CL-100
Peshawby, Albert = Pesherbay, Albert CL-100
Peshekee, Mark = CL-099
Pesherbay, Albert = CL-100
Pe-to-zo-on-oquette = Awanakwad, Petros CL-013
Pe-to-zo-ourquitte = Awanakwad, Petros CL-013
Pisherbay, Albert = Pesherbay, Albert CL-100
Poneshing, Joseph = Tazhedewin, Joseph CL-126
Po-ne'-sing = Tazhedewin, Joseph CL-126
Ponishing, Joe = Tazhedewin, Joseph CL-126
Porcellay, Charles = CL-101
Porsley, Charles = Porcellay, Charles CL-101
Prestawin, Jacob = CL-102
Quabaway, James = Quoboway, James CL-103
Quar-bo-way, James = Quoboway, James CL-103
Quoboway, James = CL-103
Recalt, S = Penaiswanquot, Jacko CL-098
Ricott, S = Penaiswanquot, Jacko CL-098
Rubberdee, John = CL-104
Sacketup, Gorge = Kabayacega, George CL-059
Sagatup = Kabayacega, George CL-059
Sahgahnahquato, Aaron = CL-105
Sanegoby, Simon = Sanequaby, Simon CL-106
Sanequaby, Simon = CL-106
Sa-Qua-On = Sawbequom, Adam CL-108
Sargarnargeeton, Aron = Sahgahnahquato CL-105
Sargaswongeto, Aaron = Sahgahnahquato CL-105
Sargon-quatt, Aaron = Sahgahnahquato CL-105
Sargonquatto, Aaron = Sahgahnahquato CL-105
Sargonquongay, Aaron = Sahgahnahquato CL-105
Sashkobanquot = CL-107
Sash-ko-ban-quot = Sashkobanquot CL-107
Saw-be-come, Adam = Sawbequom, Adam CL-108
Sawbecome, Xedan = Sawbequom, Adam CL-108
Sawbequom, Adam = CL-108
Scholes, Edward = Jackson, Edward CL-053
Schyahuahquet, Aaron = Sahgahnahquato CL-105
Scott, Antoine = CL-109
Seneyobym, Simon = Sanequaby, Simon CL-106
Seymour, Joseph = CL-110
Shabema, Charles = Shabena, Charles CL-111
Shabena, Charles = CL-111
Shake = Penaiswanquot, Jacko CL-098
Shako = Penaiswanquot, Jacko CL-098
Shanernuskin, Joseph = Shawonosang CL-115
Shaw, Charles = CL-113
Shaw-an-ax, Joseph = Shawanese, Joseph CL-114
Shawanese, Joseph = CL-114
Shaw-an-ese, Joseph = Shawanese CL-114
Shawanosang, Joseph = Shawonosang CL-115

 All Rights Reserved by Chris Czopek

ALL NAMES INDEX OF COMPANY K SOLDIERS

Shaw-au-ase, Joseph = Shawanese CL-114
Shaw-au-os-ang, Joseph = Shawonosang CL-115
Shaw-em-as-ang, Joseph = Shawonosang CL-115
Shawnes, Jos. = Shawanese, Joseph CL-114
Shawonosang, Joseph = CL-115
Shaw-wakum, Joseph = Shawonosang CL-115
Shay-Go-Nay-Be, Thomas = Smith CL-120
Shedoinktuk, Joseph = Shawanese CL-114
Shegog, John = Shegoge, John CL-116
Shegoge, John = CL-116
Shomin, John B. = CL-117
Shomin, John O. = CL-118
Shomin, Louis = CL-119
Shonin, Lewis = Shomin, Louis CL-119
Shornin, Louis = Shomin, Louis CL-119
Shosh-bono-quot = Sashkobanquot CL-107
Shouman, Louis = Shomin, Louis CL-119
Shovet, Jos = Nesogot, Joseph CL-087
Shoyat, Joseph = Nesogot, Joseph CL-087
Singoby, Simon = Sanequaby, Simon CL-106
Smith, Thomas = CL-120
Sockum, A. = Sawbequom, Adam CL-108
South, Peter = CL-121
Stgobe, Simeon = Sanequaby, Simon CL-106
Stoneman, George = CL-122
Sutton, Freeman = CL-123
Tab Zout, Antoine = Tabyant, Antoine CL-125
Tabagah, L. = Kabaosa, Louis CL-58
Tabasasch, Francis = Tabasash, Francis CL-124
Tabasash, Francis = CL-124
Tabash, Francis = Tabasash, Francis CL-124
Tabiah, Antoine = Tabyant, Antoine CL-125
Tabusgream, James = Dabasequam, Jonah CL-030
Tabusgream, James = Tabasash, Francis CL-124
Tabyant, Antoine = CL-125
Tal-ghe-de-roin, Joseph = Tazhedewin CL-126
Ta-ohe-de-niw, Joseph = Tazhedewin CL-126
Tay-bay-se-gay = David, John CL-031
Tazhedewin, Joseph = CL-126
Ta-zhe-de-win, Joseph = Tazhedewin CL-126
Tchibatis, Amable = Ketchebatis, Amable CL-066
Thomas, Moses = CL-127
Tobiah, Antoine = Tabyant, Antoine CL-125
Tobyant, Jno. = Tabyant, Antoine CL-125
Toshidowe, Joseph = Tazhedewin, Joseph CL-126
Valentine, Robert = CL-128
Valintine, Robert = Valentine, Robert CL-128
Wa Wa Chi Nodin = Newton, William CL-089
Wabana, John = Waubenoo, John CL-135
Wabano, Thomas = CL-129

ALL NAMES INDEX OF COMPANY K SOLDIERS

Wabesis, Charles = CL-130
Wabesis, John = CL-131
Wabsis, Charles = Wabesis, Charles CL-130
Wa-Bu-No, Thomas = Wabano, Thomas CL-129
Waggishegomy, William = Waygeshegoing CL-137
Wahacanhamaza, Joseph = Wakazoo, Joseph CL-132
Wah-Bah-Noo, John = Waubenoo, John CL-135
Wakago, Joseph = Wakazoo, Joseph CL-132
Wakazoo, Joseph = CL-132
Wanageshik, Big Jim = Arwonogezice CL-008
Wanbenoo, John = Waubenoo, John CL-135
Wasageshik, James = Arwonogezice CL-008
Wasakashic, James = Arwonogezice CL-008
Wa-sas-cum = Waygeshegoing, William CL-137
Wasegesik, James = Arwonogezice CL-008
Wash kah, Bavrg-man = Mashkaw, John CL-075
Washegesic, James = Arwonogezice CL-008
Wassagezhic, James = Arwonogezice CL-008
Wassagezic, Henry = CL-133
Watson, James V. = CL-134
Waubanoo, Thomas = Wabano, Thomas CL-129
Wau-bau-no, Thomas = Wabano, Thomas CL-129
Waubenoo, John = CL-135
Waubesis, Charles = Wabesis, Charles CL-130
Waukazoo, Joseph = Wakazoo, Joseph CL-132
Wau-sha-gee-sic, James = Arwonogezice CL-008
Waw-be-naw, John = Waubenoo, John CL-135
Waw-be-noo, John = Waubenoo, John CL-135
Waw-be-sis, John = Wabesis, John CL-131
Wayaubemind, Michael = CL-136
Waygeshegoing, William = CL-137
We-ash-kid-ba, Noah = Narwegeshequabey CL-084
Wegesaquina, William = Waygeshegoing CL-137
Wells, Daniel = Mwakewenah, Daniel CL-082
Wells, Peter = CL-138
Wesaw, Thomas = CL-139
Wesler, John = Wesley, John CL-140
Wesley, John = CL-140
Wessley, John = Wesley, John CL-140
Wezoo, Thomas = Wesaw, Thomas CL-139
Whitbrain, C V = Whiteface, Charles CL-141
Whiteface, Charles = CL-141
Wi yaw be mind, Antoine = Scott, Antoine CL-109
Williams, James = CL-142
Williams, Joseph = Nesogot, Joseph CL-087
Williams, Moses = CL-143
Williams, Samuel = CL-144
Wolf, Payson = CL-145
Wolfe, Paysow = Wolf, Payson CL-145
Wosegesic, Jim = Arwonogezice, James CL-008

 All Rights Reserved by Chris Czopek

COMPLETE ROSTER OF THE SOLDIERS IN COMPANY K

THE CHRIS LIST

This is my personal list of names for the soldiers of Company K.
In BOLD TYPE is my choice for what the correct name should be. Call it the "PRIME NAME".
The "aka" names are either alternate names or alternate spellings.
My only source of information are army records, therefore this list is not "final".
Someday an anishinabemowin teacher will re-write this list, using cultural heritage as a guide.
Till then, use my list to find out who are the soldiers in Company K, and where they came from.
Because there are so many different spellings of the names, I have given each soldier in Co. K a sort of "army serial number". You can call it a "Chris List Number".
Some names have more than 25 different spellings! Using a CL number brings them together.
Some soldiers had two names. Example: Aaron Sahgahnahquato - Aaron Pequongay.
In these cases, giving each name the same CL number shows the connection.

AGE - This is the age of the soldier when he enlisted in 1863-1864-1865.
HOME - The home town of the soldier at the time he enlisted, according to army records.
AKA names marked (Pension) or (Pension Name) appear on the Pension File index cards.
G.A.R. - A national veterans organization for Civil War soldiers, just like the VFW of today.
Most towns in Michigan had a G.A.R. Post. Members met once a month, wore a special uniform, marched in parades, and ordered gravestones from the government for any veteran who died.
Almost every gravestone for a Company K soldier was ordered by the local G.A.R. Post.

CHRIS LIST# **NAME** **HOME**

CL-001 **Agahgo, Charles** Home: Pentwater
aka: A-gaer-go / Chas. Agorgo (gravestone) Age: 24
DIED OF DISEASE - Buried in Alexandria National Cemetery under the name "Chas. Agorgo".

CL-002 **Allen, Charles** Home: Northport
aka: "Charles Allen Nash-kaze" Age: 19
DIED FROM BATTLE WOUNDS received in the Battle of the Wilderness. 1st Sgt. of Co. K

CL-003 **Amderling, Peter** Home: Rockland
aka: Anderson / Underling / Anotagan Age: 26
SURVIVED - He was recruited in the Upper Peninsula.

CL-004 **Anderson, Peter** Home: Mason Co.
aka: Aish-ke-baw-go-ne-quot Age: ??
Enlisted during the last days of war. Never mustered into service, never left Michigan.

 All Rights Reserved by Chris Czopek

CHRIS LIST#	NAME	HOME

CL-005 Andress, Edwin V. Home: Chesaning
aka: Andrew(s) Age: 25
OFFICER - Captain of Co. K (not Native American). Wounded in battle. Discharged for disability.

CL-006 Andrew, John Home: Isabella
aka: Jno Anderson (name on his gravestone) Age: 18
SURVIVED (WOUNDED IN ACTION) Buried in Taymouth twp., Saginaw Co. as "Jno Anderson".

CL-007 Aptargeshick, Oliver Walpole Island
aka: Ar-pe-targe-zhik Age: 19
DIED FROM WOUNDS - Buried in Arlington National Cemetery. Monument on Walpole Island.

CL-008 Arwonogezice, James Home: Pentwater
aka: Arno-woge-zice / Wassagezhic (death certificate) Age: 25
SURVIVED - Over 30 ways to spell his name. Member of the G.A.R. in Northport and in Charlevoix.

CL-009 Ashkanak, Joseph Cross Village
aka: Ach-Au-Nach / Jos Ach Aw Nach (gravestone) Age: 18
SURVIVED (WOUNDED IN ACTION) - Member of the Cross Village G.A.R. Post No.106.

CL-010 Ashkebug, George Home: Northport
aka: Ashe-ke-bug / Estabrook Age: 28
SURVIVED - Grave recently discovered on the Leelanau peninsula, in Onominise Cemetery.

CL-011 Ashkebugnekay, Amos Home: Pentwater
aka: Ash-ne-buy-nec-kay / Amos Green Age: 32
CAPTURED - P.O.W. - Survived Andersonville Prison and the Sultana steamboat explosion.

CL-012 Ashman, Daniel Home: Isabella
aka: None found Age: 17
CAPTURED - DIED WHILE P.O.W. during move from Andersonville Prison to Millen, GA.

CL-013 Awanakwad, Petros Harbor Springs
aka: Petros A-won-o-quot / Pe-to-zo-ourquitte Age: 25
KILLED BY ARTILLERY FIRE - After the war his widow received a pension from the government.

CL-014 Battice, John Home: Lansing
aka: Bottis Age: 19
SURVIVED - He was paid to substitute for a man who was going to be drafted.

CHRIS LIST#	NAME	HOME
CL-015	**Benasis, John**	Home: Algonquin
	aka: Bensus / John Bird	Age: 22

SURVIVED - RETURNED HOME He was recruited in the Upper Peninsula (Ontonagon county).

CL-016	**Bennett, Louis**	Harbor Springs
	aka: None found	Age: 31

SURVIVED - RETURNED HOME - Buried on Mackinac Island in historic St. Anne cemetery.

CL-017	**Boushaw, Augustus**	Harbor Springs
	aka: Bushaw, Beauchamp, Bemassawe, Paymosayway	Age: 20

SURVIVED - Died March 23, 1887. Buried in Harbor Springs, Lakeview Cemetery.

CL-018	**Burns, Peter**	Home: Isabella
	aka: Barnes	Age: 21

DIED FROM DISEASE - Buried in Arlington National Cemetery, Washington D.C.

CL-019	**Cabecoung, William**	Home: Isabella
	aka: Kai bai coung / Chippen or Chebbin or Chabin	Age: 19

SURVIVED - Lived in the St. Charles area (Saginaw Co.) Died at Quanicassee around 1891.

CL-020	**Carter, Charles**	Hagerstown, MD
	aka: None found	Age: 21

KILLED IN ACTION - He was a member of the Delaware Tribe in the state of Maryland.

CL-021	**Chamberlain, Amos**	Home: Isabella
	aka: None found	Age: 22

SURVIVED - Member of the Mt. Pleasant G.A.R. (WA-BU-NO Post No. 250) Died August 18, 1909.

CL-022	**Chatfield, Charles**	Home: Isabella
	aka: None found	Age: 19

DIED FROM DISEASE – Was wounded in action on June 17, 1864.

CL-023	**Chatfield, Samuel C**	Home: Isabella
	aka: "W. Samuel Chatfield" / Chalfield (pension name)	Age: 26

KILLED IN ACTION during the Siege of Petersburg, VA.

CL-024	**Church, Albert**	St. Charles
	aka: "James Church" (gravestone) / Chirch (pension)	Age: 20

SURVIVED - Gravestone in St. Charles Riverside cemetery says "James Church".

CHRIS LIST#	NAME	HOME
CL-025	**Collins, Jacob**	Home: Isabella
	aka: None found	Age: 30
	SURVIVED - WOUNDED IN ACTION at "The Crater", Petersburg, VA.	
CL-026	**Collins, John**	Home: Isabella
	aka: None found	Age: 21
	SURVIVED - During the war he was both a soldier and a hospital attendant.	
CL-027	**Collins, William**	Home: Isabella
	aka: None found	Age: 33
	FATE UNCERTAIN - NEEDS TO BE RESEARCHED	
CL-028	**Corbin, George**	Home: Isabella
	aka: None found	Age: 23
	SURVIVED - Member of Mt Pleasant G.A.R. Post 250. Died May 20, 1928. He was 97 years old.	
CL-029	**Crane, Amos**	Home: Baraga Co.
	aka: None found	Age: 28
	SURVIVED – From the U.P., Paid to be a substitute for DeWitt C. Leach, the Indian Commissioner.	
CL-030	**Dabasequam, Jonah**	Harbor Springs
	aka: "Jonah Dabasequam" / Labosquam	Age: 20
	KILLED IN ACTION at the battle of Spotsylvania, VA.	
CL-031	**David, John**	Home: Isabella
	aka: "Jno Davis" (name on gravestone) / Tabaseka	Age: 21
	DIED OF TYPHUS FEVER Buried in Annapolis National Cemetery as "Jno Davis".	
CL-032	**DeLand, James S.**	Home: Jackson
	aka: "James J. DeLand"	Age: 28
	OFFICER - Captain of Co. K (not Native American). Brother of regiment commander, Col. DeLand.	
CL-033	**Driggs, William J.**	East Saginaw
	aka: None found	Age: 23
	OFFICER - 1st Lieutenant of Co. K (not Native American). Son of a Michigan Congressman.	
CL-034	**Dutton, Luke**	Home: Taymouth
	aka: "Luther Dutton" / Dutton, Luther	Age: 19
	SURVIVED - Wounded in action. Died in 1870. Buried in Taymouth twp., Saginaw County.	

CHRIS LIST#	NAME	HOME

CL-035 Etarwegeshig, John — Home: Pentwater

aka: E-tar-ivegezhik / Etawegazhik / Waw-me-gwan — Age: 26

KILLED IN ACTION - After the war his mother, Tah-te-go-quah Etarwegeeshig, received a pension.

CL-036 Genereau Jr., Louis — Home: Pentwater

aka: "Louis Genereau" / "Louis Genereaux" — Age: 18

SURVIVED - Wounded in action. Son of a well known interpreter. Died June 25, 1871 at age 27.

CL-037 George, David — Home: Isabella

aka: None (son of George and Nancy Nagatoe) — Age: 22

DIED FROM BATTLE WOUNDS Buried in Arlington National Cemetery, Washington D.C.

CL-038 Gibson, Joseph L. — Home: Bear Creek

aka: "Joseph Na-baw-na-ya-sang" / "Louis J. Gibson" — Age: 23

CAPTURED - DIED WHILE P.O.W. in Andersonville Prison - Buried there in grave # 7741.

CL-039 Going, Samuel — Home: Pentwater

aka: Goeng (pension name) / "Samuel Gora" (gravestone) — Age: 20

DIED FROM WOUNDS received at Battle of Spotsylvania, VA. Joined the army on 4th of July.

CL-040 Graveraet, Garrett A. — Harbor Springs

aka: Gaveraet (wrong spelling on army records) — Age: 23

DIED FROM WOUNDS Officer - 2nd Lieutenant (Native American). Grave on Mackinac Island.

CL-041 Graveraet Jr., Henry G. — Harbor Springs

aka: None found — Age: 55

DIED FROM BATTLE WOUNDS. 1st Sgt. of Co. K. Buried at Fredericksburg National Cemetery.

CL-042 Greensky, Benjamin C. — Home: Charlevoix

aka: None found — Age: 24

KILLED IN ACTION His father was minister at Greensky Hill church, Charlevoix. Brother of Jacob.

CL-043 Greensky, Jacob — Home: Charlevoix

aka: None found — Age: 18

SURVIVED - Helped build the log church on Greensky Hill, Charlevoix. Brother of Benjamin.

CL-044 Gruet, Peter — Home: Marquette

aka: Gruett / Peter Grewett Jr. — Age: 21

SURVIVED – After enlisting in Co. K, he enlisted in 9th Mich. Cavalry

CHRIS LIST#	NAME	HOME

CL-045 Hall, Cornelius Home: Ludington

aka: Ne-gaw-ne-say Age: 18

SURVIVED - Died August 5, 1917. Buried west of Houghton Lake in Star City Cemetery.

CL-046 Hamlin, James H Home: St. Ignace

aka: "James M. Hamblin" Age: 43

CAPTURED - DIED WHILE P.O.W. in Andersonville Prison - buried there in grave # 11,260.

CL-047 Hannin, Joseph Home: Canada

aka: Hannins / Hannius / Honyust (Death Certificate) Age: 25

SURVIVED – Iroquois of the Oneida tribe. Lived on Oneida Reserve 41 in Middlesex County, Canada.

CL-048 Hubert, Charles Sault Ste. Marie

aka: Herbert / "C. Hubbard" (on hospital records) Age: 28

WOUNDED IN ACTION - Sent to hospital in Philadelphia. No further record.

CL-049 Isaacs, John Home: Isabella

aka: None found Age: 21

SURVIVED - RETURNED HOME. Brother of William Isaacs.

CL-050 Isaacs, William Home: Isabella

aka: Isaac (name on pension file index card) Age: 18

SURVIVED (WOUNDED IN ACTION) - His Civil War discharge paper was saved by his family.

CL-051 Jacko, John Home: Northport

aka: "John Shawkoo" (Durant Roll) Age: 20

SURVIVED - Son of Jacko Penaiswonquot, a Co.K soldier who died at Andersonville Prison.

CL-052 Jacko, Nattahmenoting Home: Northport

aka: Jackonat-bah-me-no-ling / "Louis Natawinoding" Age: 21

SURVIVED - Died after the war from disease. Buried in Detroit Elmwood Cemetery.

CL-053 Jackson, Edward Andrew Home: Canada

aka: "Andrew Jackson" is the name on his army records. Age: 27

SURVIVED - Born in New York state. Died Dec 23, 1893. Buried in Canada (Blenheim, Ontario).

CL-054 Jackson, William Home: Saginaw

aka: None found Age: 26

SURVIVED - RETURNED HOME Paid to be a substitute for a drafted soldier.

CHRIS LIST#	NAME	HOME
CL-055	**Jeandron, Michael**	Home: St. Ignace
	aka: "M. Jourdron" (gravestone) / Judson / Jondrau	Age: 21

CAPTURED - DIED WHILE P.O.W. in Andersonville Prison - buried there in grave # 12,396.

CL-056	**Johns, David**	Home: Kawkawlin
	aka: "David John" (pension) / "David Waw-bo-ge-maw"	Age: 19

SURVIVED – Chippewa from Bay City area. Brother of William Johns. Died in 1889 in Mason County.

CL-057	**Johns, William**	Home: Kawkawlin
	aka: "William Waw-bo-ge-maw" / Jack, William	Age: 20

SURVIVED – Chippewa from Bay City area. Brother of David Johns.

CL-058	**Kabaosa, Louis**	Home: Pentwater
	aka: "Lewis Kubaosa"	Age: 27

SURVIVED - Joined the army on the 4th of July at Elbridge (Oceana County).

CL-059	**Kabayacega, George**	Home: Saginaw
	aka: Kabayahsega (gravestone) / "George Sacketup"	Age: 30

SURVIVED - He was a paid substitute for a man who was going to be drafted.

CL-060	**Kadah, Joseph**	Home: Pentwater
	aka: Hagler (gravestone) / Heogler	Age: 21

DIED OF DISEASE 1864 in St. Mary's Hospital, Detroit. Buried under the name "Joseph Hagler".

CL-061	**Kakakee, Joseph**	Home: Chester
	aka: Me-She-Ka-Kack / Kukakee	Age: 35

SURVIVED - Name means "Pigeon Hawk". Chester is in Ottawa Co. Buried in Mason County.

CL-062	**Kaquatch, Samuel**	Harbor Springs
	aka: Kagwaich (pension file) / Kagwaitch (gravestone)	Age: 18

DIED IN THE WAR Wounded In Action - Killed in a war-time train accident.

CL-063	**Kechittigo, Thomas**	East Saginaw
	aka: "Thomas K. Chetego"	Age: 27

SURVIVED - Lived in Grayling. Member of the Grayling G.A.R. (Marvin Post No. 240).

CL-064	**Keeses, John**	Home: Richland
	aka: John Kesis	Age: 30

SURVIVED - Died Oct 20, 1867 at Wayland, Michigan. Richland is in Kalamazoo County.

CHRIS LIST#	NAME	HOME

CL-065 Kejikowe, Simon — Home: Burt Lake

aka: Kigegoway / Gijigowi — Age: 23

TWICE WOUNDED. Sent to a hospital and never heard from again. Member of Burt Lake band.

CL-066 Ketchebatis, Amable — Cross Village

aka: Kitcherbalist / Tchibatis / "A. Chibadice" — Age: 32

DIED WHILE P.O.W. - Captured during fight at "The Crater". Died in the prison camp at Danville, VA.

CL-067 Kewaconda, Benjamin — Home: St. Ignace

aka: Kowuaiconda — Age: 21

SURVIVED – Came from St. Ignace on the Straits of Mackinac. Occupation: Boatman.

CL-068 Kinewahwanipi, John — Leelanau twp.

aka: Kiniewahasoipi / Ke-new-nuh-ne-ba / Penegwah / Pashabe — Age: 20

DIED OF DISEASE - Joined the army in Grand Rapids. Died three weeks later in Virginia.

CL-069 Lamourandere, Thaddeus — Home: Pentwater

aka: "Taddy Lamorander" / Lamourandieux — Age: 21

KILLED IN ACTION at Spotsylvania, VA.. After the war, his father received a pension.

CL-070 Lidger, David — Home: Mendon

aka: Lidyer / Leidyer — Age: 28

SURVIVED – Iroquois of the Oneida tribe, born in Oneida, NY. Enlisted together with Joseph Hannin.

CL-071 Light, Josiah — Home: Pentwater

aka: None found — Age: 18

SURVIVED - RETURNED HOME - Fought at "The Crater" Petersburg, VA.

CL-072 Marks, Louis — Home: Petoskey

aka: None (son of Enos Marks of Petoskey) — Age: 24

CAPTURED - P.O.W. He survived Andersonville Prison and returned home after the war.

CL-073 Marquette, Frank — Home: Assyria

aka: "Francis Marquette" — Age: 35

SURVIVED - A professional hunter who became an army sharpshooter. Assyria is in Barry County.

CL-074 Mashkaw, James — Home: Pentwater

aka: Kaw-gaw-ge-way — Age: 31

KILLED IN ACTION He and his brother joined the army together, and they died in battle together.

CHRIS LIST#	NAME	HOME
CL-075	**Mashkaw, John**	Home: Pentwater
	aka: Marcum / Mash-kah	Age: 22

KILLED IN ACTION He and his brother joined the army together, and they died in battle together.

CL-076	**Miller, Thomas**	Home: Omena
	aka: Kah-gee / Kahgee	Age: 19

DIED OF DISEASE during the war while home on convalescent leave. Buried in Omena, Michigan.

CL-077	**Misisaius, Edward**	Cross Village
	aka: Misisins / Minissisance	Age: 25

DIED IN THE WAR – Wounded in battle, sent to a hospital, and was never heard from again.

CL-078	**Miskoguon, Louis**	Home: Charlevoix
	aka: Muskoguon / "Lewis Miscogeon" (pension)	Age: 30

CAPTURED - P.O.W. - Survived Andersonville Prison and the Sultana steamboat explosion.

CL-079	**Mixinasaw, William**	Home: Northport
	aka: "W. Mackswaser" (gravestone) / Mesenasaw	Age: 28

CAPTURED - DIED WHILE P.O.W. in Andersonville Prison - buried there in grave # 11,511.

CL-080	**Mixonauby, Thomas**	Home: Pentwater
	aka: Mixonanby	Age: 39

DIED OF DISEASE in a Detroit Hospital on March 12, 1865.

CL-081	**Mogage, George W.**	Athens twp.
	aka: Moogargoe (gravestone and pension name)	Age: 23

WOUNDED IN ACTION - Died in 1923. Buried in the Grand Rapids Veterans Home cemetery.

CL-082	**Mwakewenah, Daniel**	Home: Petoskey
	aka: "Dan'l Mokewenaw" (gravestone) / "Daniel Wells"	Age: 40

DIED FROM WOUNDS - Chief of the Bear River band. Buried in Brady Hill Cemetery, Saginaw.

CL-083	**Narquaquot, Joseph**	Home: Pentwater
	aka: Naroquaquot	Age: 25

DIED OF DISEASE in Chicago, while guarding Confederate prisoners at Camp Douglas.

CL-084	**Narwegeshequabey, Jackson**	Home: Isabella
	aka: "Noah We-ash-kid-ba", Nar-ue-ge-she-gwa-by	Age: 22

MISSING IN ACTION at Battle of the Crater, Petersburg, VA on July 30, 1864. No further record.

 All Rights Reserved by Chris Czopek

CHRIS LIST#	NAME	HOME

CL-085 **Nauquam, Thompson** Home: Charlevoix

aka: Nauquam Age: 23

FATE UNCERTAIN - NEEDS TO BE RESEARCHED

CL-086 **Nelson, Thomas** Home: Taymouth

aka: None found Age: 27

SURVIVED (WOUNDED IN ACTION) - Buried in Taymouth twp., Saginaw County.

CL-087 **Nesogot, Joseph** Home: Isabella

aka: Joseph Williams Age: 27

CAPTURED - P.O.W. - Survived Andersonville Prison. Buried in Taymouth twp., Saginaw County.

CL-088 **Neveaux, William** La Pointe, Wisc.

aka: Nevarre / Nevaue Age: 18

SURVIVED - He came from Madeline Island, Wisconsin, the spiritual home of the Ojibwe people.

CL-089 **Newton, William** Home: Pentwater

aka: "William C. Knowlton" (name in obituary) Age: 17

CAPTURED - P.O.W. - Survived Andersonville Prison. Died in 1922. Buried in Yuba cemetery.

CL-090 **Nichols, Lemuel R.** Fayette twp.

aka: None found Age: 21

OFFICER - 1st Lieutenant of Co. K (not Native American). Fayette township is in Hillsdale County.

CL-091 **Noyes, Kirke W.** South Haven

aka: None found Age: 32

OFFICER - 1st Lieutenant of Co. K (not Native American). Died Aug 8, 1920, in South Haven.

CL-092 **Ohbowakemo, James** Ionia County

aka: "James R. Oshburn" / Aubwakoum Age: 26

SURVIVED - RETURNED HOME - He died in 1895 in Nottawa township, Isabella County.

CL-093 **Otashquabono, Leon** Harbor Springs

aka: "Leo Odesheweiad" (name in obituary) Age: 23

SURVIVED - Member of the Charlevoix G.A.R. (Baxter Post No.119). Buried in Harbor Springs.

CL-094 **Otto, Marcus** Home: Isabella

aka: Ke-ne-we-ge-shik / Kenewegoshik Age: 19

DISCHARGED as a DISABLED VETERAN. Lost an arm at the battle of Spotsylvania, VA.

CHRIS LIST#	NAME	HOME
CL-095	**Otto, Solomon**	Home: Isabella

aka: Kargayarsang, Solomon — Age: 22

DIED IN THE WAR - Brother of Marcus Otto.

CL-096	**Pakemaboga, John**	Home: Unknown

aka: Pa-ke-nabo-ga — Age: ??

KILLED IN ACTION at Petersburg, VA.

CL-097	**Pemassegay, Daniel**	Home: Isabella

aka: None found — Age: 24

FATE UNCERTAIN - NEEDS TO BE RESEARCHED

CL-098	**Penaiswanquot, Jacko**	Home: Northport

aka: "Jacko Pe-nais-now-o-quot" / "S. Recalt" — Age: 40

CAPTURED - DIED WHILE P.O.W. in Andersonville Prison - buried in grave # 11,705 (S. Recalt).

CL-099	**Peshekee, Mark**	Home: Isabella

aka: None found — Age: 30

SURVIVED - His name means “buffalo”. Wounded in the left shoulder at the battle of Spotsylvania.

CL-100	**Pesherbay, Albert**	Home: Charlevoix

aka: Pisherbay / Pershabay (name in obituary) — Age: 25

SURVIVED - Died in 1903. Buried in an unmarked grave in Greenwood Cemetery, Petoskey.

CL-101	**Porcellay, Charles**	Home: Burt Lake

aka: Parceley / Porsley — Age: 25

SEVERELY WOUNDED - Discharged as a disabled soldier.

CL-102	**Prestawin, Jacob**	Home: Bingham

aka: None found — Age: 24

WOUNDED Sent to a hospital. No further record. His hometown, Bingham, is in Leelanau County.

CL-103	**Quoboway, James**	Home: Isabella

aka: Quabaway / Quar-bo-way — Age: 32

SURVIVED (WOUNDED IN ACTION)

CL-104	**Rubberdee, John**	Home: Rockland

aka: Rabberden — Age: 25

SURVIVED - RETURNED HOME - He was recruited in the Upper Peninsula.

CHRIS LIST#	NAME	HOME
CL-105	**Sahgahnahquato, Aaron**	Leelanau twp.
	aka: Sargonquatto / "Aaron Pequongay"	Age: 21

SURVIVED - RETURNED HOME - Member of Northport G.A.R. Post No.399. Buried in Omena.

CL-106	**Sanequaby, Simon**	Home: Burt Lake
	aka: Singoby / Sinegoby / Say-nin-gwaw-day (Durant Roll)	Age: 25

SURVIVED - Member of the Burt Lake band. Member of Cross Village G.A.R. Post No.106.

CL-107	**Sashkobanquot**	Home: Pinconning
	aka: "Shash Kobanquot" (pension name) / Shosh-bono-quot	Age: 45

DIED IN THE WAR in an accident while guarding Confederate prisoners at Camp Douglas, Chicago.

CL-108	**Sawbequom, Adam**	Home: Pentwater
	aka: Saw-be-come / "A. Sockum"	Age: 28

CAPTURED - DIED WHILE P.O.W. in Andersonville Prison - buried there in grave # 11,508.

CL-109	**Scott, Antoine**	Home: Pentwater
	aka: Antoine Le Croix / Antoine Wi-yaw-be-mind	Age: 22

SURVIVED - ** Hero - Recommended to receive the Medal of Honor ** (But it was not awarded).

CL-110	**Seymour, Joseph**	Home: Greenland
	aka: None found	Age: 20

SURVIVED - RETURNED HOME He was recruited in the Upper Peninsula (Ontonagon county).

CL-111	**Shabena, Charles**	Cross Village
	aka: Shabema, Charles	Age: 23

SURVIVED - RETURNED HOME - He enlisted as a wagon driver for Company K.

CL-112	**Shanahan, Winfield S**	La Grange twp.
	aka: None found	Age: 18

OFFICER - 2nd Lieutenant of Co. K (not Native American) Transferred to Co. K last months of war.

CL-113	**Shaw, Charles**	Home: Pentwater
	aka: None found	Age: 19

SURVIVED - Was a member of the regiment's Color Guard. Buried in Hart, Michigan (Oceana Co.).

CL-114	**Shawanese, Joseph**	Home: Chesaning
	aka: Shaw-au-ase / "Jos Shawnes" (gravestone)	Age: 23

SURVIVED - RETURNED HOME - Buried in St. Charles Riverside Cemetery as "Jos. Shawnes".

CHRIS LIST#	NAME	HOME
CL-115	**Shawonosang, Joseph**	Beaver Island
	aka: Shaw-au-os-ang / Shaner-nuskin	Age: 18

SURVIVED (SEVERELY WOUNDED) Before the war & after the war, he lived on Beaver Island.

CL-116	**Shegoge, John**	Home: Pentwater
	aka: Shegog / Sheyoy	Age: 21

DISCHARGED as a DISABLED VETERAN - Lost an arm in the fighting at Petersburg, VA.

CL-117	**Shomin, John B.**	Cross Village
	aka: "John B.S. Shomin" / Shoman	Age: 20

CAPTURED - P.O.W. - Survived Andersonville Prison. Died in 1913. Buried on Mackinac Island.

CL-118	**Shomin, John O.**	Cross Village
	aka: Shamin	Age: 18

SURVIVED (WOUNDED IN ACTION) Died July 3, 1895 at Harbor Springs. Buried Lakeview Cemetery.

CL-119	**Shomin, Louis**	Cross Village
	aka: Shouman (gravestone) / Shor-nin (pension name)	Age: 19

SURVIVED - Member of the Cross Village G.A.R. Post No.106. Brother of John B. Shomin.

CL-120	**Smith, Thomas**	Home: Isabella
	aka: Shay-Go-Nay-Be, Thomas	Age: 23

SURVIVED - RETURNED HOME - Promoted to First Sergeant of Company K. Blinded by disease.

CL-121	**South, Peter**	Home: Petoskey
	aka: None found	Age: 18

CAPTURED - DIED WHILE P.O.W. in Andersonville Prison - buried there in grave # 12,310.

CL-122	**Stoneman, George**	Home: Pentwater
	aka: None found	Age: 22

SURVIVED (WOUNDED IN ACTION) - Joined the army on the 4th of July at Elbridge.

CL-123	**Sutton, Freeman**	Home: Ludington
	aka: None found	Age: 32

SURVIVED - Wounded in the leg during the final assault on Petersburg, VA.

CL-124	**Tabasash, Francis**	Cross Village
	aka: Tabasasch	Age: 38

SURVIVED - Promoted to Sergeant. Charter member of the Cross Village G.A.R. Post No.106.

CHRIS LIST#	NAME	HOME
CL-125	**Tabyant, Antoine**	Harbor Springs

aka: "Jno Tobyant" (gravestone name) / Tobiah — Age: 35

SURVIVED (WOUNDED IN ACTION) Buried in Harbor Springs. Gravestone says "Jno Tobyant".

CL-126 **Tazhedewin, Joseph** — Home: Pentwater

aka: Ta-ohe-de-niw / Poneshing — Age: 19

SURVIVED - WOUNDED IN ACTION at Petersburg, VA . Died Jan, 3, 1923, near Elbridge

CL-127 **Thomas, Moses** — Home: Pinconning

aka: "Moses, Thomas" (= error on army records) — Age: 18

SURVIVED - RETURNED HOME – Born in Bay County, Died in Canada 1933 - Last survivor of Co. K.

CL-128 **Valentine, Robert** — Home: Lansing

aka: Valintine (spelling on pension file index card) — Age: 23

SURVIVED - Paid to be a substitute for a drafted soldier. Died in 1902 or 1903 in Wisconsin.

CL-129 **Wabano, Thomas** — Home: Isabella

aka: Wau-bau-no / Wa-Bu-No (gravestone) — Age: 24

DIED OF DISEASE - Mt. Pleasant's G.A.R. was named after this man (WA-BU-NO Post No.250).

CL-130 **Wabesis, Charles** — Home: Pentwater

aka: Waubesis / Wabsis — Age: 25

SURVIVED - Joined the army on the 4th of July at Elbridge (Oceana County).

CL-131 **Wabesis, John** — Home: Pentwater

aka: Wa-be-sis — Age: 23

KILLED IN ACTION while fighting at "The Crater", Petersburg, VA, on July 30, 1864.

CL-132 **Wakazoo, Joseph** — Grand Traverse Bay

aka: Waukazoo / Waukazo — Age: 23

SURVIVED - First he enlisted in 16th Mich. Infantry. Later he transferred to Co. K, 1st Mich. SS.

CL-133 **Wassagezic, Henry** — Home: Pewabic

aka: Wassagezie / "Henry Condeacon" — Age: 17

SURVIVED - Recruited in the Upper Peninsula. He died April 16, 1922 at Odanah, Wisconsin.

CL-134 **Watson, James V.** — Home: Pentwater

aka: None found — Age: 44

DISCHARGED FOR DISABILITY - Joined the army at Age 44.

CHRIS LIST#	NAME	HOME
CL-135	**Waubenoo, John**	Home: Isabella
	aka: Wah-bah-noo / Wanbenoo	Age: 22

SURVIVED - Died March 25, 1910. Buried near Weidman in Fairview cemetery (Isabella County).

CL-136	**Wayaubemind, Michael**	Home: Petoskey
	aka: Mitchell Wayohlamin	Age: 19

SURVIVED (TWICE WOUNDED IN ACTION). Enlisted at Bear River, the old name for Petoskey.

CL-137	**Waygeshegoing, William**	Home: Isabella
	aka: Waggishegomy (pension name) / Gisecomey	Age: 22

SURVIVED - RETURNED HOME - After the war, he lived in the St. Charles area of Saginaw County.

CL-138	**Wells, Peter**	Home: Petoskey
	aka: None found	Age: 18

SURVIVED - RETURNED HOME - Enlisted at "Bear River". Later, name was changed to "Petoskey".

CL-139	**Wesaw, Thomas**	Athens twp.
	aka: Wezoo (gravestone and pension name)	Age: 22

SURVIVED (WOUNDED IN ACTION) Lived on the Nottawaseppi Huron Reservation in Calhoun Co.

CL-140	**Wesley, John**	Oceana County
	aka: Wesler / Wessley / Naw-yah-waw-sung (Durant Roll)	Age: 28

SURVIVED - RETURNED HOME - Died Sept 26, 1908 in Grant township, Mason County.

CL-141	**Whiteface, Charles**	Home: Pentwater
	aka: "C.V. Whitbrain" (gravestone) / Now-quo-um	Age: 22

DIED OF DISEASE in Chicago, while guarding Confederate prisoners at Camp Douglas.

CL-142	**Williams, James**	Home: Lapeer
	aka: None found	Age: 18

FATE UNCERTAIN - NEEDS TO BE RESEARCHED

CL-143	**Williams, Moses**	La Pointe, Wisc.
	aka: None found	Age: 30

KILLED IN ACTION at "The Crater" - Was from La Pointe, Wisconsin. Enlisted in Marquette, Michigan.

CL-144	**Williams, Samuel**	Home: Lapeer
	aka: None found	Age: 20

FATE UNCERTAIN - NEEDS TO BE RESEARCHED

CHRIS LIST# **NAME** **HOME**

CL-145 **Wolf, Payson** Home: Northport

aka: Wolfe, Paysow Age: 30

CAPTURED - P.O.W. - Survived Andersonville Prison - Member of Cross Village G.A.R. Post 106.

CL-146 **Kedgnal, John** Grand Haven

aka: Kegnot / "John Henry Duvernay" (gravestone) Age: 18

CAPTURED - P.O.W. - Survived Salisbury Prison - Native American in Company B (not Co. K).

CL-147 **Duvernay, William** Grand Haven

aka: Duverney / "Chid Duverney" Age: 13

SURVIVED Native American drummer boy in Company B, age 13 (not a member of Company K).

Photo by Chris Czopek

CHARLES AGAHGO 001

A-Gaer-Go, Charles, Pentwater. Enlisted in company K, First Sharpshooters, July 4, 1863, at Elbridge, for 3 years, age 24. Mustered July 11, 1863. Died of disease at Alexandria, Va., Jan. 10, 1865.

Agahgo, Charles – PRIME NAME, from Pension File
A-Gaer-Go, Charles – Adjutant General Record name
Agorgo, Charles – Gravestone name
Agaho, Tah-gutch-quay – 1883 Pension Roll (widow)
Agago, Charles – 1890 Veteran Census / Index book (widow)
Pawgawchequay – Widow's name Durant Roll S-Durant Roll Field Notes #7-pg50

RESIDENCE IN 1863: Elbridge township, Oceana County S-AGR

DEATH DATE: January 19, 1865 S-pf-001 (pension file for Charles Agahgo)
(Died of Disease in an Army Hospital)

GRAVE: Alexandria National Cemetery in Alexandria, VA. / **Section B, Grave #3391**

PENSION FILE: Yes, see Appendix A
1883 Pension Roll: Yes, Oceana Co. – Hart (Widow)

VETERAN CENSUS: 1888 No / 1890 Yes(widow) / 1894 No
RAY HEREK's BOOK: Pages: 415, 434(Roster)

SPECIAL INFORMATION:
His name means "Old Porcupine". S-ccn001a
Charles Agahgo was a Grand River Ottawa living on the reservation at Elbridge near Pentwater. He enlisted in Company K on the 4th of July. Two days later he said goodbye to his wife and children at the dock in Pentwater, and went off to war on a steamboat. That was the last time his family saw him. "Old Porcupine' died of disease (enysipelas) at L'Overture General Hospital in Alexandria, VA. S-pf-001 (pension file for Charles Agahgo)
His widow, **Pah-gutch-quay**, received a widow's pension till her death in 1914 – almost 50 years. There is a Veterans Monument in Hart, Michigan (Oceana County) with his name on it.

NOTE TO HISTORIANS:
The Adjutant General Record says he died Jan 10, 1865. However, his pension file clearly says that he died January 19, 1865. Trust the pension file in these matters.
The widow's name has many spellings: On 1883 Pension Roll, Tah-gutch-quay Agaho; on the 1890 Veteran Census, Lucy Pequechquay; on the pension file index card, Pahquah Equa Agahgo; on the Durant Roll, Paw-gaw-che-quay (Durant Roll Number 5451, Field Notes #7-pg50).

CHARLES ALLEN 002

Allen, Charles, Northport. Enlisted in company K, First Sharpshooters, as Sergeant, June 12, 1863, at Northport, for 3 years, age 19. Mustered June 22, 1863. First Sergeant. Died May 18, 1864, of wounds received in action at the Wilderness, Va., May 6, 1864.

ALLEN, Charles – PRIME NAME, all sources agree on spelling

RESIDENCE IN 1863: Northport, Leelanau County S-AGR
DEATH DATE: May 18, 1864 (*Died from wounds received in battle*) S-AGR
GRAVE: Fredericksburg National Cemetery, VA (*exact location of grave is unknown, but all facts indicate he died on May 18, 1864 at the hospital in Marye House, Fredericksburg, VA, and is buried among the 13,000 unidentified soldiers in Fredericksburg National Cemetery*)

PENSION FILE: Yes, see Appendix A; 1883 Pension Roll: No

VETERAN CENSUS: 1888 No / 1890 No / 1894 No

RAY HEREK's BOOK: Pages: 34, 123, 386, 434(Roster)

SPECIAL INFORMATION:
Charles Allen enlisted as a Sergeant when he was 19 years old. Promoted to First Sergeant. S-AGR
He was mortally wounded in Company K's first battle, The Wilderness, on May 6, 1864.
This makes him the first soldier in Company K to fall on the field of battle. S-ccn002a
In Traverse City there is a Veterans Park on 11th Street and Elmwood Avenue. In this park is a block of stone with the names of local men who died in the Civil War. "Charles Allen" is one of the names on this memorial.
Charles Allen grew up on the Leelanau peninsula near Omena. His family numbered among the first members of the Rev. Dougherty's "New Mission", and Charles was probably educated at the Grove Hill school. His father had two names, Moses Nashkaze and Moses Allen. The last name "Allen" was given to Moses by the Rev. Dougherty when he was baptized, and all his children used "Allen" as their last name. The Presbyterian church in Omena was the family church. The diary of Rev. George N. Smith (Missionary in Northport) mentions Charles Allen many times. S-ccn002b
After the war, Charles' sister, Susan Allen, married Company K soldier Aaron Sahgahnahquato.

NOTE TO HISTORIANS:
The grave of Charles Allen is most likely in Fredericksburg National Cemetery for the following reasons: The diary of the Rev. George N. Smith has an entry on June 6, 1864, saying Charles Allen died in a hospital. The men of Company K wounded at the Wilderness and Spotsylvania were sent to Marye House, a hospital in Fredericksburg, VA. That is where Charles probably died.
The men who died in the Marye House hospital were temporarily buried on the property. After the war, all the graves of soldiers who died in the hospitals around Fredericksburg were moved to the nearby Fredericksburg National Cemetery. Over 13,000 unidentified graves were marked with a simple stone that says "U.S. Soldier". Charles Allen is buried in one of those graves.

PETER AMDERLING 003

Anderson, Peter. Enlisted in company K, First Sharpshooters, Sept. 26, 1864, at Rockland, for 3 years, age 26. Mustered Oct. 20, 1864. Mustered out at Delaney House, D. C., July 28, 1865.

Amderling, Peter – PRIME NAME, from Pension File
Anotagan, Peter – Pension File, aka name
Underling, Peter – Adjutant General Annual Report
Anderson, Peter – Adj. Gen. Record (Clerical error, should be Amderling, Peter)

RESIDENCE IN 1864: Rockland, Ontonagon County (Upper Peninsula) S-AGR

DEATH DATE: February 1872 *(died in Houghton County in the UP)* S-pf-003

GRAVE: Location Unknown
(*Note: since he died in Houghton County in the UP, he is probably buried there*)

PENSION FILE: Yes, see Appendix A; 1883 Pension Roll: No

VETERAN CENSUS: 1888 No / 1890 No / 1894 No

RAY HEREK's BOOK: Pages: 434(Roster, spelled Peter Anderson)

SPECIAL INFORMATION:
Company K suffered heavy losses in the summer of 1864. Messages were sent to Michigan – We need more men! Recruiters went to the Upper Peninsula in search of replacements. **Peter Amderling answered the call. He enlisted at Rockland in Ontonagon County,** then traveled to Virginia to join the regiment near Petersburg.

NOTE TO HISTORIANS:
I don't know how it happened, but long ago, the Adjutant General's office misspelled the name of Peter Amderling. They called him Peter Anderson by mistake, and nobody noticed.
For proof of this error, compare facts in the pension file for Peter Amderling to the facts in the Adjutant General Record for Peter Anderson.
Also, notice the Adjutant General Annual Report has the name "Peter Underling". Still misspelled, but close enough to Amderling to tell us, the name "Anderson" on the Adj. Gen. Record is an error.

Some people believe that Peter Amderling died in Elk Rapids, Michigan, in 1928 under the name Peter Anderson. This is not true. The man who died in Elk Rapids was a Grand River Ottawa. Peter Amderling was from the Upper Peninsula. They were two different people.
Two men enlisted in Company K from Rockland on the same day, Amderling and Rubberdee.

PETER ANDERSON 004

JANUARY 26, 1928

DIES AT 102 YEARS

Mr. Peter Anderson died at the home of his nephew, Ed. Anderson, last Saturday night at the advanced age of one hundred two years. He was born at Battle Point on the Grand Haven River. He enlisted in the United States Army in 1865 but saw no service for his company got only to Grand Rapids on its southbound journey. Rev. Mathews read the funeral service and Mr. John Dockery, undertaker, prepared the body and directed the funeral.

Obituary from the Elk Rapids Progress January 26, 1928, Front Page.
(Elk Rapids is near Traverse City, in Antrim County)

Note: His name is Peter Anderson, but the details of his army service do not match the Adj. General Record shown below. I believe these are two different men.

Anderson, Peter. Enlisted in company K, First Sharpshooters, Sept. 26, 1864, at Rockland, for 3 years, age 26. Mustered Oct. 20, 1864. Mustered out at Delaney House, D. C., July 28, 1865.

Anderson, Peter – PRIME NAME, **Obituary in the Elk Rapids Progress 1-26-1928**
Anderson, Jno – Name on gravestone for John Andrews (see below)
Aish-ke-baw-go-ne-quot – Name on Durant Roll S-Durant Field Notes #13-pg55

DEATH DATE: January 21, 1928 (*Died in Elk Rapids, Michigan*) S-County Death Certificate
GRAVE: Location Unknown (*Probably the Kewadin Indian Mission cemetery* S-ccn004a)
PENSION FILE: No **VETERAN CENSUS:** 1888 No / 1890 Yes / 1894 No
RAY HEREK's BOOK: No mention of Peter Anderson, the Grand River Ottawa

SPECIAL INFORMATION:
At the end of the Civil War, an Anishinabe named Peter Anderson enlisted in Company K, but it was too late. The war ended before he left Michigan to join his unit in Virginia.
He did not serve one day of active duty. That is the true story of Peter Anderson.
Genealogy records show he was a Grand River Ottawa, son of Shawsawteas Anderson. S-ccn004b
He appears on the 1890 Veteran Census, living in Ludington, Mason County. No dates of enlistment or discharge given, just the comment "Discharge papers lost".
Grand Traverse Herald, Feb 10, 1865, says Payson Wolf left Northport with "three Indian recruits" for Company K. Adjutant General Records show that one was John Kinewahwanipi and another was John Jacko. But the identity of the third man is a mystery – perhaps he was Peter Anderson?
NOTES FOR HISTORIANS: The Adjutant General Records show a soldier named **Peter Anderson** from Rockland in Ontonagon County. But this is not the same man who died in Elk Rapids in 1928. The soldier from Rockand is **Peter Amderling**. Somehow his name was misspelled "Anderson" in the Adjutant General Records, and nobody corrected the mistake.
There is a gravestone in Taymouth, Saginaw County, for "Jno Anderson – Co. K". This is NOT the grave of Peter Anderson. It belongs to another man in Co. K named John Andrew.

EDWIN V. ANDRESS 005

Andress, Edwin V., Chesaning. Entered service in company K, First Sharpshooters, at organization, as Captain, June 23, 1863, at Dearborn, for 3 years, age 25. Commissioned July 22, 1863. Mustered July 22, 1863. Discharged for disability July 26, 1864, on account of wounds received in action at Spottsylvania, Va., May 12, 1864.

Andress, Edwin V. - PRIME NAME, from Adj. General Record & Pension File
Andrews, Edwin – Common misspelling of his name

RESIDENCE IN 1863: Chesaning, Saginaw County S-AGR
In 1880s: Weldon, California (Kern County)

DEATH DATE: October 10, 1884 (Died near Weldon, Kern Co., California S-ccn005a)

GRAVE: California, Kern County, Onyx / Cottage Grove Cemetery
Note: There is a Memorial Inscription carved on the obelisk stone in the Andress family plot in Wildwood Cemetery, Chesaning, Saginaw County, Michigan

PENSION FILE: Yes, see Appendix A
1883 Pension Roll: Yes, California, Kern Co., Onyx (gsw r.hand & r.foot)

VETERAN CENSUS: 1888 No / 1890 Yes (widow, in Michigan) / 1894 No

RAY HEREK's BOOK: Pages: 34, 35, 36, 96, 157, 202, 367, 368, 388, 408, 434(Roster), 475, 476, 482

SPECIAL INFORMATION:
White officer (Captain) in Company K. He was the company commander.
This man was from Chesaning in Saginaw County. He was a white man, but somewhere in his past he learned to speak the Anishinabe language fluently. He understood and respected Anishinabe customs and traditions. By all indications, he got along well with the soldiers under his command. In fact, Payson Wolf named one of his sons **Edwin Andress Wolf** after him.

NOTES FOR HISTORIANS:
Andress was not wounded on May 12, he was wounded in the right foot on May 9 during the fighting at the Ni River, VA. S-Herek pg 388
Be aware, the name Andress is often written Andrews. You will come across this error frequently when you are reading papers in a pension file.
There is a famous picture of wounded soldiers from Company K sitting under a tree at Marye's Heights, near Fredericksburg, VA. The man sitting with his back against the tree and his right foot elevated & wrapped in bandages is definitely Captain Andress. S-ccn005b

JOHN ANDREW 006

Andrew, John, Isabella. Enlisted in company K, First Sharpshooters, May 18, 1863, at Isabella, for 3 years, age 18. Mustered May 26, 1863. Wounded June 17, 1864. Discharged at Detroit, Mich., Aug. 11, 1865.

Andrew, John – PRIME NAME, from Pension File
Anderson, Jno – Gravestone name (see note below)

RESIDENCE IN 1863: Isabella, Isabella County S-AGR

DEATH DATE: January 19, 1884
S-GENDIS, Dept. of Health Death Index, Ledger page 75, Record No. 276

GRAVE: Saginaw Co., Taymouth twp. / Taymouth Indian Cemetery (Military Stone)
(This cemetery is also called Wheaton Indian cemetery)
Name on the gravestone is "Jno Anderson" (see note below)

PENSION FILE: Yes, see Appendix A
1883 Pension Roll: No

VETERAN CENSUS: 1888 No / 1890 No / 1894 No

RAY HEREK's BOOK: Pages: 182, 205, 211, 396, 400, 434(Roster)

SPECIAL INFORMATION:
During the Civil War, this man was severely wounded near Petersburg, VA, during the fighting around Shand House on June 17. A bullet tore through his right thigh near the hip. S-Herek pg 182
He was in the hospital system for more than a year (June 1864 to August 1865). He, John Andrew, Augustus Boushaw, Thomas Smith, and Robert Valentine were discharged from the army together in Detroit on August 11, 1865. They were the last men in Company K to leave the service and return home.
John Andrew's wound did not heal properly, and it bothered him for the rest of his life. S-pf-006
His gravestone in the Taymouth Indian cemetery has confused people for many years. The name on it is Jno Anderson, leading people to think it belongs to a Company K soldier named Peter Anderson.

OLIVER APTARGESHICK 007

Ar-pe-targe-zhik, Oliver, Walpole Island. Enlisted in company K, First Sharpshooters, June 17, 1863, at Dearborn, for 3 years, age 19. Mustered June 17, 1863. Died at Washington, D. C., July 9, 1864, of wounds received in action at Petersburg, Va., June 17, 1864. Buried in National Cemetery at Arlington, Va.

Aptargeshick, Oliver – PRIME NAME, from gravestone
Ar-pe-targe-zhik, Oliver – Adjutant General Record name
Ahreotergeshig, O – Roll Of Honor, 1868 S-ccn007a

RESIDENCE IN 1863: Walpole Island, Canada (Lambton County, Ontario) S-AGR

DEATH DATE: July 9, 1864 (*Died from battle wounds in an army hospital*)

GRAVE: Arlington National Cemetery, Washington D.C., Section 13, # 5732

PENSION FILE: No Pension File exists

VETERAN CENSUS: 1888 No / 1890 No / 1894 No

RAY HEREK's BOOK: Pages: 395, 434(Roster)

SPECIAL INFORMATION:
He came from Walpole Island, Canada. Enlisted in Company K on June 17, 1863. Exactly one year later, on June 17, 1864, he was mortally wounded in battle. Sent to Lincoln hospital in Washington, D.C., where he died July 9. **Buried in Arlington National Cemetery – Section 13, Grave #5732.** He never returned to Walpole Island, but the people there have not forgotten him. The veterans on the island have made a memorial to his memory near the monument to Chief Tecumseh. It is a block of granite with his name, the crossed rifles of an infantryman, and the eagle feathers of an Anishinabe warrior. This memorial to a Company K soldier is little known and seldom visited.

NOTES FOR TRAVELERS AND HISTORIANS:
Walpole Island is located on the border between Michigan and Canada. You can find it on a map near Detroit, on the north-east shore of Lake St. Clair. It belongs to Canada, and is part of Lambton County, Ontario. Easiest way to get there is by ferry boat from Algonac, Michigan. (This is a border crossing, so remember to bring your passport or a Michigan Enhanced Driver's License).
Walpole Island is an Indian Reservation, home to the Walpole Island First Nation.
A good time to visit is during the annual Pow Wow held on the third Saturday of July.

JAMES ARWONOGEZICE 008

Arno-woge-zice, James, Pentwater. Enlisted in company K, First Sharpshooters, Sept. 12, 1863, at Little Traverse, for 3 years, age 25. Mustered Oct. 21, 1863. Discharged at Detroit, Mich., Aug. 11, 1865. Present residence, Northport, Mich.

Arwonogezice, James – PRIME NAME, from 1894 Veteran Census
Arno-woge-zice, James – Adjutant General Record name
Arwonogezia, James – Pension File name
Wassagezhic, James – Death Certificate (County clerk office, Charlevoix Co.)
Wasagesick, James – 1890 Veteran Census (he is on this census twice)
Washegesic, James – Veterans Memorial on Beaver Island
Wau-sha-gee-sic, James – Newspaper story in 1905 S-ccn008a
Lightsky, Jim – Newspaper obituary S-Charlevoix Courier, May 23, 1917

RESIDENCE IN 1863: Pentwater, Oceana County S-AGR
From 1865-1898, Charlevoix; 1899-1900, Omena & Northport; 1901, Elk Rapids; From 1902-07, Veterans Home, Grand Rapids; 1907-1917, Charlevoix; S-ccn008b
DEATH DATE: May 17, 1917 (*Died at Bay Shore, Hayes twp., Charlevoix Co.*) S-Death Cert.
GRAVE: Charlevoix County, Charlevoix / Brookside Cemetery S-State Death Record
PENSION FILE: Yes, see Appendix A; 1883 Pension Roll: No
VETERAN CENSUS: 1888 Yes / 1890 Yes / 1894 Yes
RAY HEREK's BOOK: Pages: 434(Roster)

SPECIAL INFORMATION:
James Arwonogezice often used "Wassagezhic" as his name (spelled many different ways). To prove this is true, compare the information on his Civil War pension file index card under the name "James Arwonogezia" with his death certificate in the Charlevoix County Clerk's office under the name "James Wassageshic". Notice how both men have the first name James and both died May 17, 1917, in Bay Shore. Without a doubt, they are the same person.
Notice there was another soldier in Company K with a similar last name, Henry Wassagezic. James lived in Charlevoix, Henry lived in Wisconsin.
In 1886 there was a reunion of the 1st Mich. Sharpshooters in Jackson. James attended this reunion. His name appears in the reunion booklet as "Was-a-ge-sik, Jas…Charlevoix". S-ccn008c
A photograph of "Big Jim Wanageshik" c. 1905 can be seen in the museum in Northport, MI.
Location of his grave is a problem. One obituary says he was buried near Bay Shore, another says Greensky Hill, and his State Death Record says Brookside Cemetery, City of Charlevoix. S-ccn008d
The Grand Rapids Veterans Facility was started as an "Old Soldiers Home" for Civil War veterans. James lived there five years. Find him in the Soldiers Home Record Book as resident # 3742.
James was member #75 of Baxter G.A.R. Post No. 119, Charlevoix, Charlevoix County.
He was also a member of Woolsey G.A.R. Post No. 399, Northport, Leelanau County.
On Beaver Island there is a Veterans Memorial; his name is on a brick as "James Washegesic".

JOSEPH ASHKANAK 009

Ach-Au-Nach, Joseph, La Croix. Enlisted in company K, First Sharpshooters, July 14, 1863, at Traverse, for 3 years, age 18. Mustered July 20, 1863. Discharged at Detroit, Mich., Aug. 11, 1865.

Ashkanak, Joseph – PRIME NAME, from Pension File
Ach-Au-Nach, Joseph – Adjutant General Record name
Ach Aw Nach, Jos – Gravestone name
Tabash, Joseph – 1888 Veteran Census

RESIDENCE IN 1863: La Croix (=Cross Village), Emmet County S-AGR
In 1890: Cross Village, Emmet County S-1890 Veteran Census

DEATH DATE: May 31, 1914 (*Died at Cross Village, Emmet Co.*) S-State Death Record

GRAVE: Emmet County, Cross Village / Holy Cross Cemetery (Military stone)
(*Location: Holy Cross Cemetery is on Levering road, exactly one mile east of Cross Village*)

PENSION FILE: Yes, see Appendix A
1883 Pension Roll: Yes, Emmet Co. - Cross Village (gsw left arm)
(Note: gsw = gun shot wound)

VETERAN CENSUS: 1888 Yes / 1890 Yes / 1894 Yes

RAY HEREK's BOOK: Pages: 372, 434(Roster)

SPECIAL INFORMATION:
Joseph Ashkanak was from "La Croix", a place known today as **Cross Village** in Emmet County. This area has a special place in Michigan history. It is one of the oldest settlements in the state, thanks to the large Anishinabe population living there which attracted missionaries like Father Marquette and Bishop Baraga. **Eight men from Cross Village volunteered for Company K.** Joseph Ashkanak was one of those men. He was wounded in battle – gun shot wound, left arm. When the war was over, the veterans got together and formed an organization called the G.A.R. **Joseph was a charter member of the Washington G.A.R. Post No. 106, Cross Village. The G.A.R. was like the VFW of today. They met once a month, marched in parades on holidays, put flags on veteran graves, and lobbied congress for better veteran benefits.**

NOTES FOR HISTORIANS:
Adjutant General Record says Joseph was discharged August 11, 1865, which is correct.
The Adjutant General Annual Report says he was discharged Aug. 21, 1865, which is an error.
The 1888 Veteran Census gives his name as "Joseph Tabash". Maybe someone thought he was the brother of Francis Tabasash, who lived nearby and is listed on the census as "Francis Tabash".

GEORGE ASHKEBUG 010

Ashe-ke-bug, George, Northport. Enlisted in company K, First Sharpshooters, July 9, 1863, at Northport, for 3 years, age 28. Mustered July 20, 1863. Mustered out at Delaney House, D. C., July 28, 1865.

Ashkebug, George – PRIME NAME, from Pension File
Ashe-ke-bug, George – Adjutant General Record name
Estabrook, George – Pension File, aka name
Askabcook, George – Pension File, aka name
Azh-ke-buck, George – Name on 1872 Land Patent S-ccn010a
Ashe-ke-bugasberg, George – History of Leelanau Township, Pg 61

RESIDENCE IN 1863: Northport, Leelanau County S-AGR

DEATH DATE: July 23, 1886 S-Shier, pg 59

GRAVE: Leelanau County / Onominese Cemetery (Military stone)
(*Note: In my opinion, the iron G.A.R. flag holder that stands alone in the center of this cemetery marks the correct location of his grave, and the gravestone was originally beside it* S-ccn010b)

PENSION FILE: Yes, see Appendix A
1883 Pension Roll: No

VETERAN CENSUS: 1888 No / 1890 No / 1894 No

RAY HEREK's BOOK: Pages: 434(Roster)

SPECIAL INFORMATION: At the time of his death, George Ashkebug did not have a decent set of clothes to be buried in. His brother in law mortgaged a team of horses to buy George a suit for the funeral. S-pf-010 pension file for George Ashkebug

NOTE FOR HISTORIANS:
Onominese Cemetery is the last remnant of "Onumunese Village", a small settlement located on the west side of Leelanau peninsula near Gills Pier. To find it on a map, imagine a line running straight west from Northport. Where that line meets Lake Michigan is where the village stood. It is often mentioned in the diary of Rev. George N. Smith. For many years, the cemetery was forgotten. Today, the Grand Traverse Band of Ottawa and Chippewa Indians in Peshawbestown is taking care of it.

AMOS ASHKEBUGNEKAY 011

Ash-ne-buy-nec-kay, Amos, Pentwater. Enlisted in company K, First Sharpshooters, July 4, 1863, at Pentwater, for 3 years, age 32. Mustered July 11, 1863. Missing in action at Petersburg, Va., June 17, 1864. Discharged at Detroit, Mich., Aug. 1, 1865.

Ashkebugnekay, Amos – PRIME NAME, from Pension File
Ash-ne-buy-nec-kay, Amos – Adjutant General Record
Ash-bug-ne-ka, Amos – 1890 Veteran Census
Green, Amos – Name he used after the war, on all records
Green, Arnold – Andersonville National Park website
Green, Philip – Name in his obituary S-Scottville Enterprise, June 28, 1906, Front Page
Greenarm, Amos – aka name, appears in a Pension File S-pf-001 (Agahgo)
Ish-ka-buga-ma-ka – See Herek's book pg 281 for an interesting story

RESIDENCE IN 1863: Elbridge, Oceana County S-AGR
After the war: He lived near Weldon Creek, Custer twp., Mason Co. S-1890 Vet. Census
In 1894: He lived 2 miles East, and ½ mile South of Custer, Michigan S-pf-051 (Jacko)
DEATH DATE: June 19, 1906 S-State Death Record
GRAVE: Mason Co., Custer twp., Custer /Custer Riverside Cemetery (Unmarked)
Grave is unmarked, and exact location of his grave inside the cemetery is unknown
(*The town of Custer is east of Ludington on US-10; the cemetery is south of town on Custer Road*)
PENSION FILE: Yes, see Appendix A; 1883 Pension Roll: No
VETERAN CENSUS: 1888 Yes / 1890 Yes / 1894 No
RAY HEREK's BOOK: Pages: 398, 434(Roster)

SPECIAL INFORMATION:
The service record of this Civil War soldier is awesome. He survived disease, the biggest killer of soldiers, the bloody Battle of Spotsylvania, nine months at Andersonville prison, and the Sultana tragedy. The Sultana was a steamboat hired to pick up recently released POWs and take them home. A boiler exploded one night and the ship caught fire. Of an estimated 2,400 passengers, only 600-800 survived. Amos Ashkebugnekay was one of those few survivors. His name appears on a list of Sultana survivors published in a Detroit newspaper. S-ccn011a
After the war, he used Amos Green as his name and lived a quiet life as a farmer.
When he died in 1906, the Scottville newspaper said : "Mr. Green not only had the respect of his own people, but of every white man who had the pleasure of his acquaintance".
S-Scottville Enterprise, June 28, 1906, Front Page
He joined the Haight G.A.R. Post No. 348 in Scottville, Mason Co., as "Amos Green". S-ccn011b

QUESTION FOR HISTORIANS:
Is there a group picture of the Scottville G.A.R. Post No. 348 in existence that shows Amos Green standing in the ranks?

DANIEL ASHMAN 012

Ashman, Daniel, Isabella. Enlisted in company K, First Sharpshooters, as Corporal, May 18, 1863, at Isabella, for 3 years, age 17. Mustered May 26, 1863. Taken prisoner while charging the enemy's works before Petersburg, Va., June 17, 1864. Died in prison at Andersonville, Ga.

Ashman, Daniel – PRIME NAME, all records agree

RESIDENCE IN 1863: Isabella, Isabella County S-AGR

DEATH DATE: Shortly after November 11, 1864 (*Died while a POW*)

GRAVE: Location Unknown
Note: He did not die at Andersonville prison, but at another place, probably at Camp Lawton, the prison camp at Millen, GA S-Research by Lansing historian Roger Winthrop)

PENSION FILE: No Pension File exists

VETERAN CENSUS: 1888 No / 1890 No / 1894 No

RAY HEREK's BOOK: Pages: 398, 413, 434(Roster)

SPECIAL INFORMATION:
Daniel Ashman must have been an outstanding person. When he joined the army at age 17 he was immediately made a Corporal.
One source says he was a Drummer. S-Fancher's "Past and Present, Isabella County", pg 79.
Unfortunately, he was one of the 15 soldiers from Company K captured at Petersburg, VA., on June 17, 1864 (same fight where Lt. Graveraet was mortally wounded) and sent to the infamous Andersonville Prison in Georgia. Official Army records say he died there.
Recently, Lansing historian Roger Winthrop has carefully investigated the circumstances of his death. He discovered that Daniel Ashman left Andersonville during a transfer of prisoners to a prison camp near Millen, Georgia, on Nov 11, 1864. No further record. It appears Daniel died either on the train to Millen, or shortly after arrival at the Millen prison camp.

PETROS AWANAKWAD 013

A-won-o-quot, Petros, Little Traverse. Enlisted in company K, First Sharpshooters, June 8, 1863, at Little Traverse, for 3 years, age 25. Mustered June 20, 1863. Died of wounds received in action Feb. 20, 1865.

Pe-to-zo-ourquitte. Member of company K, First Sharpshooters. Died March 29, 1865. Buried near Petersburg, Va.

Awanakwad, Petros – PRIME NAME, from Pension File
A-won-o-quot, Petros – Adjutant General Record name
Pe-to-zo-ourquitte – Adjutant General Record name, misspelled
Awanakwad, Betois – 1890 Veteran Census / Index book (widow)

RESIDENCE IN 1863: Little Traverse (=Harbor Springs), Emmet County S-AGR
DEATH DATE: February 20, 1865 S-AGR
GRAVE: Poplar Grove National Cemetery, near Petersburg, VA., Grave #0561
Name on cemetery records is Pe-to-zo-on-oquette. The name on his gravestone at Poplar Grove Cemetery is Jaco Penaswonquot, which is completely wrong.
PENSION FILE: Yes, see Appendix A
1883 Pension Roll: Yes, Emmet Co. - Cross Village (Widow)
VETERAN CENSUS: 1888 No / 1890 Yes (widow) / 1894 No
RAY HEREK's BOOK: Pages: 406, 434(Roster)
Petozoourquitte is mentioned in Ray Herek's book on pages 313, 418, 435(Roster)

SPECIAL INFORMATION:
Last soldier in Company K to die on a battlefield. He was hit by shell fire and his right leg was shattered. He died in a field hospital a few hours later. Buried in Poplar Grove National Cemetery. He left behind a wife in Middle Village named Teresa Awanakwad and two young children, Bazile and Benjamin. Teresa collected a widow's pension for 45 years. She died Oct 19, 1910 and is buried in the Catholic cemetery in Good Hart. Her death record has the name "Thresa Bidos".
ERROR IN OFFICIAL RECORDS: The Adjutant General Record has a soldier named "Pe-to-zo-ourquitte", no first name, on the roster of Company K, with no information other than date and place of death. There is no doubt that Petozoourquitte = Petros Awanakwad.
The record for Pe-to-zo-ourquitte says the date of his death is March 29, 1865. That is an error. March 29 is the day his body was reburied in Poplar Grove National Cemetery. S-ccn013a

NOTE TO HISTORIANS: His first name was Peter. It seems the missionary who baptized him used the Greek version "Petros". Family and friends called this soldier Betoes or Betos, because Native Americans sometimes pronounce the letter "P" to sound like the letter "B", and they have difficulty pronouncing the letter "R". Some records call him Pierre or Pease, confirming that his first name was Peter. When his widow died in 1910, the name put on her death certificate was Thresa Bidos. It appears she used her husband's first name as her last name (Bidos = Betos/Betoes). One son, Benjamin Betos, likewise used his father's first name as his last name.

JOHN BATTICE 014

Battice, John. Enlisted in company K, First Sharpshooters, Aug. 24, 1864, at Jackson, for 3 years, age 19. Mustered Aug. 24, 1864. Substitute for Daniel Vanakin. Mustered out at Delaney House, D. C., July 28, 1865.

Battice, John – PRIME NAME, Adjutant General Record name
Bottis, John – Adjutant General Annual Report S-ccn014a

RESIDENCE IN 1864: Enlisted in Jackson, Michigan S-AGR
Credited to Lansing for draft quota purposes S-Descriptive Roll, State Archives
John Battice was recruited in the Upper Peninsula S-Herek pg 250

DEATH DATE: Unknown

GRAVE: Location Unknown

PENSION FILE: No Pension File exists

VETERAN CENSUS: 1888 No / 1890 No / 1894 No

RAY HEREK's BOOK: Pages: 250, 434(Roster)

SPECIAL INFORMATION:
He enlisted in Jackson, Michigan, as a substitute for Daniel Vanakin. S-AGR
It was common for a man who was being drafted to pay someone to take his place.
John Battice probably received a hundred dollars for doing this, big money in those days.
According to Herek's book, page 250, **John Battice was recruited in the Upper Peninsula.**
The reason why Lansing receives credit for him in the draft is a matter for speculation.

JOHN BENASIS 015

Bensus, or Benasis, John. Enlisted in company K, First Sharpshooters, Oct. 10, 1864, at Algonquin, for 3 years, age 22. Mustered Oct. 20, 1864. Joined regiment at Peebles' Farm, Va., Nov. 4, 1864. Mustered out at Delaney House, D. C., July 28, 1865.

Benasis, John – PRIME NAME, Adjutant General Record name
Bensus, John – Adjutant General Record, aka name
Bennesis, John – Adjutant General Annual Report
Bird, John – 1890 & 1894 Veteran Census name

RESIDENCE IN 1864: Algonquin, Ontonagon County S-Descriptive Roll, State Archives
1882 – 1907: In Baraga County (in the U.P.), by the historic Kewawenon Mission (aka "the Methodist Mission") at Zeba, near L'Anse S-1890 Veteran Census & his obituary

DEATH DATE: May 9, 1907 S-State Death Record for "John Bird"

GRAVE: Baraga County, L'Anse / The Pinery Indian Cemetery S-State Death Record
(There is no marker on his grave, and no record of where inside the cemetery he is buried. His death record says he was buried in the Methodist Mission Cemetery, which is another name for the Pinery Indian Cemetery)

PENSION FILE: No Pension File exists

VETERAN CENSUS: 1888 No / 1890 Yes / 1894 Yes

RAY HEREK's BOOK: Pages: 434(Roster)

SPECIAL INFORMATION:
His name means "bird". At the end of his life, friends and neighbors called him "John Bird". John Benasis came from Michigan's Upper Peninsula. He was probably a member of the historic L'Anse band of the Lake Superior Chippewa, which today is known as the Keweenaw Bay Indian Community. He had strong ties with the Kewawenon Mission near L'Anse, in Baraga County. When he returned home after the war, he changed his name to "John Bird", lived at the mission, and worked as a lay preacher. S-Kalamazoo Valley Family Newsletter # 7, Winter 1977 / 78 Pg 110 (Vol. 7), article titled "Indians of Ontonagon County in Company K, First Michigan Sharpshooters".

The Adjutant General Record gives two spellings of his name: Bensus and Benasis. I think Benasis sounds more like an Anishinabe name, so I chose it for the Prime Name. He was recruited in the U.P. late in the war, served nine months, and was honorably discharged with the rest of the regiment. His obituary appeared on the front page of the L'Anse Sentinel, May 11, 1907. It mentions that "John Bird" was an Indian and a veteran of the Civil War. He was buried in the Pinery Indian Cemetery. The grave is unmarked and the location unknown.

LOUIS BENNETT 016

Bennett, Louis, Little Traverse. Enlisted in company K, First Sharpshooters, June 12, 1863, at Little Traverse, for 3 years, age 31. Mustered June 22, 1863. Mustered out at Delaney House, D. C., July 28, 1865.

Bennett, Louis – PRIME NAME, all records agree

RESIDENCES:

In 1832: Mackinac Island (born there in March, 1832) S-Military Record, National Archives
In 1863: Little Traverse (=Harbor Springs), Emmet County S-Descriptive Rolls, State Archives
In 1869: Mackinac Island (Died in Holmes twp., Mackinac County) S-US Federal Census Mortality Schedule, 1870

DEATH DATE: November 19, 1869 S-Shier pg 76
GRAVE: Mackinac Co., Mackinac Island / St. Anne Cemetery (Military stone)
PENSION FILE: Yes, see Appendix A; 1883 Pension Roll: No
VETERAN CENSUS: 1888 No / 1890 No / 1894 No
RAY HEREK's BOOK: Pages: 338, 434(Roster)

SPECIAL INFORMATION:
He was 31 when he joined the army, expecting to be sent into battle to fight for his country. When the regiment was sent to Chicago to guard prisoners, he was deeply disappointed. Records show he deserted for a while. But after the Sharpshooters were sent to Virginia to fight in battles, Louis Bennett traveled to Virginia and rejoined the regiment. The army recognized his Native American background and decided to go easy on him, saying he was a "Chippeway, and at the time of his desertion was unacquainted with the rules and customs of the U.S. Service". All was forgiven. At the end of the war Louis received an honorable discharge. S-Herek pg 338
Louis Bennett is buried on Mackinac Island in historic St. Anne cemetery.

NOTES FOR HISTORIANS:
The military record of Louis Bennett at National Archives give these details of his life:
He was born on Mackinac Island in March of 1832.
At the time that Louis enlisted in Company K, he was a farmer. Location of his farm is uncertain, but his enlistment in Harbor Springs indicates it was in Emmet County. S-army enlistment paper
The Federal Census Mortality Schedule for 1870 says Louis Bennet died on Mackinac Island in Holmes township in November of 1869. Cause of death: Inflammation of pleura (lung disease). His grave on Mackinac Island is marked by a military gravestone. Years ago it was forgotten, dirty, moss covered, and tilted. Today the stone stands straight and clean.

Louis Bennett was married to Rachel St.Andrew. They had one child, a daughter named Mary Jane. She was born July 16, 1866, in Tawas City. She married a man from England named Joseph Coles on June 26, 1889, at St. Ignace. No children. Mary Jane applied for a pension from the government in 1891 using her married name "Mary J. Coles", but did not receive it. She died in Kingsley, MI on August 15, 1914. Buried in Mancelona, MI. S-obituary in the Mancelona Herald Aug 20, 1914.

AUGUSTUS BOUSHAW 017

Bushaw, Augustus, Little Traverse. Enlisted in company K, First Sharpshooters, June 28, 1863, at Little Traverse, for 3 years, age 20. Mustered July 20, 1863. Discharged at Detroit, Mich., Aug. 11, 1865.

Boushaw, Augustus – PRIME NAME, the name on his gravestone
Bushaw, Augustus – Adjutant General Record
"Gus Bushaw" – Obituary S-Newspaper: Charlevoix Sentinel, April 6, 1887, Pg4 Col 2
Beauchamp, Augustus – 1860 US Census
Bushais, Augustes – 1880 US Census
Be-mas-sa-we / Pay-mo-say-way S-Marriage record / Durant Roll Field Notes #17-pg32

RESIDENCES:
In 1842: Born on the Grand River near Grand Rapids in Kent County (See note below)
In 1860: Emmet County near Harbor Springs S-1860 US Census
In 1880: Living on a farm in Friendship township, Emmet County S-1880 US Census

DEATH DATE: March 23, 1887 S-Herek pg 367
GRAVE: Emmet County, Harbor Springs / Lakeview Cemetery (Military stone)

PENSION FILE: No Pension File exists
VETERAN CENSUS: 1888 No / 1890 No / 1894 No
RAY HEREK's BOOK: Pages: xii, 367, 434(Roster)

SPECIAL INFORMATION:
His friends called him "Gus". He was the son of Charles Beauchamp, a fur trader who established a trading post on the Grand River and married a Native girl. His son Augustus was born where Grand Rapids is today. S-Birth Record of Catharine, daughter of Augustus
After the treaty of 1855, Charles moved his family to Little Traverse (today called Harbor Springs) and ran a store there. Gus appears on the 1860 US Census as Augustus Beauchamp. His Anishinabe name was Be-mas-sa-we / Pay-mo-say-way. S-Marriage record / S-Durant Roll Field Notes #17-pg32
Gus married Lucy Wasson on February 5, 1862. S-Emmet County Marriage Records
When Lt. Garrett Graveraet came to Harbor Springs in 1863 to recruit for Company K, Gus left his young wife and baby to fight for his country.
At the end of the war, he was in a hospital in Detroit recovering from either sickness or a wound. Soon as he was well enough to travel home, he was given a discharge. Census records tell us he returned to his wife in Harbor Springs and settled down on a farm in Friendship township, Emmet County. S-1880 US Census as "Augustes Bushais".
He died in 1887 and is buried in Lakeview Cemetery, Harbor Springs. His military gravestone is decorated with a flag every year on Veterans Day.

PETER BURNS 018

Burns, or Barnes, Peter, Isabella. Enlisted in company K, First Sharpshooters, May 10, 1863, at East Saginaw, for 3 years, age 21. Mustered June 16, 1863. Sergeant. Died Sept. 14, 1864. Buried in National Cemetery at Arlington, Va.

Burns, Peter – PRIME NAME, Adjutant General Record name
Barnes, Peter – Adjutant General Record, aka name

RESIDENCE IN 1863: Isabella, Isabella County S-AGR

DEATH DATE: September 14, 1864 (*Died in an army hospital in Washington D.C.*)

GRAVE: Arlington National Cemetery, Washington D.C., Section 13, # 8141

PENSION FILE: No Pension File exists

VETERAN CENSUS: 1888 No / 1890 No / 1894 No

RAY HEREK's BOOK: Pages: 415, 434(Roster)

SPECIAL INFORMATION:
Peters Burns of 1st Michigan Sharpshooters, Co. K, died from disease on September 14, 1864, in Douglas Hospital, Washington D.C., and is buried in Arlington National Cemetery. My source for this information is in State Archives of Michigan, 59-14 Ovs 84 Folder 11, Document 39, in "Report of Michigan Soldiers who have died in Hospitals, 1862-1865".

NOTE TO HISTORIANS:
There are two soldiers named Peter Burns buried in Arlington National Cemetery in Section 13. Both men died at about the same time, and both were from a Company K. Grave number 8141 is for the Native American from Company K, 1st Michigan Sharpshooters. Grave number 7500 is for a New Jersey man who served in the 1st New Jersey Cavalry, Company K.

WILLIAM CABECOUNG 019

Ca-be-coung, William, Isabella. Enlisted in company K, First Sharpshooters, May 18, 1863, at Isabella, for 3 years, age 19. Mustered May 26, 1863. Corporal May 1, 1865. Mustered out at Delaney House, D. C., July 28, 1865.

Ca-be-coung, William – PRIME NAME, Adjutant General Record
Cay-bar-coung, William – Pension File, aka name
Cowing, William C. – Pension File, aka name
Cowing Cayba, William – Pension File, aka name
Kai-bai-cong, William – Pension File, aka name S-pf-059 (Kabayacega pension file)
Chippen or Chebbin or Chabin – Pension File, aka names
Cobecong, William – Pension File, aka name

RESIDENCE IN 1863: Isabella, Isabella County S-AGR
After the war: St. Charles / Taymouth, Saginaw County S-pf-019
In 1891: Quanicassee, Tuscola County (near Bay City) S-pf-019

DEATH DATE: August 7, 1891 S-book Totems and Trees, by Dorothy L. Dinsmore
(Died at Quanicassee, Wisner township, Tuscola County) S-pf-019

GRAVE: Location Unknown

PENSION FILE: Yes, see Appendix A
1883 Pension Roll: No

VETERAN CENSUS: 1888 No / 1890 No / 1894 No

RAY HEREK's BOOK: Pages: 434(Roster)

SPECIAL INFORMATION:
The name Cabecoung probably means "waterfall". (The Ojibwe called Niagra Falls "Kichi-Kabekong" meaning "Great Falls". S-Mishomis Book, by Edward Benton-Banai, pg. 98).
William Cabecoung was also called "Chippen". S-pf-019, testimony of Chief of the Taymouth band.
NOTES FOR HISTORIANS:
Quanicassee was a Native American fishing village long before the arrival of white settlers.
This is where William Cabecoung spent the last years of his life.
The pension file of George Kabayacega contains testimony given by William Cabecoung.

CHARLES CARTER 020

Carter, Charles. Enlisted in company K, First Sharpshooters, June 22, 1864, at Petersburg, Va., for 3 years, age 21. Mustered June 22, 1864. Killed before Petersburg, Va., July 3, 1864.

Carter, Charles – PRIME NAME, all sources have same name.

RESIDENCE IN 1863: Hagerstown, Maryland S-Descriptive Rolls, State Archives

DEATH DATE: July 30, 1864 (*KIA Battle of the Crater, Petersburg, VA* S-Herek pg 402)
Note: KIA = Killed in action.

GRAVE: Location Unknown

PENSION FILE: No Pension File exists

VETERAN CENSUS: 1888 No / 1890 No / 1894 No

RAY HEREK's BOOK: Pages: 398, 434(Roster)

SPECIAL INFORMATION:
Killed in action at the Battle of the Crater, Petersburg, VA. S-Herek pg 402
This soldier was from a tribe in the state of Maryland.
The tribes of the Great Lakes region share a common language and tradition with some tribes on the East Coast. Apparently, Charles Carter came from one of those East Coast tribes.
Writer Elizabeth Edwards indicates he came from the Delaware Tribe (Read her story in TRAVERSE MAGAZINE, February 2003 issue, titled "The Bravery Of Company K".)
The Descriptive Roll in State Archives says his home was Hagerstown, Maryland.
Charles Carter joined Company K in Virginia during the siege of Petersburg.
Six weeks later, he was killed in action at the Battle of the Crater.

NOTE TO HISTORIANS:
There is an error in the official records. The Adjutant General Record says he died July 3. This is wrong. Actual date of death was July 30. S-Herek pg 402

AMOS CHAMBERLAIN 021

Chamberlain, Amos, Isabella. Enlisted in company K, First Sharpshooters, June 13, 1863, at East Saginaw, for 3 years, age 22. Mustered June 16, 1863. Corporal May 1, 1865. Mustered out at Delaney House, D. C., July 28, 1865.

Chamberlain, Amos – PRIME NAME, Adjutant General Record
Chamberlin, Amos – 1894 Veteran Census name

RESIDENCE IN 1863: Isabella, Isabella County S-AGR
In 1888: Delwin, Denver township, Isabella County S-pf-059 (Kabayacega pension file)

In 1894: Denver township, Isabella County S-1894 Veteran Census

DEATH DATE: August 18, 1909 S-pf-021

GRAVE: Location Unknown (*Probably in Isabella County, Denver township*)

PENSION FILE: Yes, see Appendix A
1883 Pension Roll: No

VETERAN CENSUS: 1888 No / 1890 No / 1894 Yes

RAY HEREK's BOOK: Pages: 434(Roster)

SPECIAL INFORMATION:
He joined Company K as a Private. Was promoted to Corporal.
In 1891, Amos joined the G.A.R., a veterans organization similar to the VFW of today. He was member #160 of the Wa-Bu-No G.A.R. Post No. 250, Mt. Pleasant, Isabella Co.
Note: This G.A.R. Post was named after Thomas Wabano of Company K. It was the only G.A.R. Post in the country named after a Native American soldier.

CHARLES CHATFIELD 022

Chatfield, Charles, Isabella. Enlisted in company K, First Sharpshooters, May 18, 1863, at Isabella, for 3 years, age 19. Mustered May 26, 1863. Absent sick Sept., 1864. No further record.

Chatfield, Charles – PRIME NAME, from Adjutant General Record

RESIDENCE IN 1863: Isabella, Isabella County S-AGR
DEATH DATE: Unknown
GRAVE: Location Unknown
PENSION FILE: No pension file exists (Note: There is a pension file for a soldier named Charles Chatfield, 10th Mich. Inf., Co. B. See Appendix A. Also note, the roster of the 10th Michigan Infantry does not have a Charles Chatfield. Perhaps the pension file belongs to the Charles Chatfield in Co. K of 1 Mich SS ?)

VETERAN CENSUS: 1888 No / 1890 No / 1894 No

RAY HEREK's BOOK: Pages: 396, 434(Roster)

SPECIAL INFORMATION:
Charles Chatfield, age 19, enlisted on the same day & same place as Samuel C. Chatfield, age 26. They were probably related (cousins?) and decided to join the army together.
Charles Chatfield was wounded in the back on June 17, 1864 at Petersburg, VA S-Herek pg 396.
He was sent to a hospital. No one ever heard from him again. His exact fate is a mystery.

The mystery of Charles Chatfield has been solved, according to genealogist Quita V. Shier. She believes that Charles Chatfield survived the war, returned home, lived in Isabella County, and died there on August 18, 1909. Complete details of his life after he left the army can be found on pages 89-92 of her book "Warriors In Mr. Lincoln's Army".

NOTE TO HISTORIANS:
The Descriptive Roll at State Archives says "8/18/09 Diseased per letter from Mount Pleasant". At face value, this note indicates that in 1909 the Adjutant General Office received a letter about him, probably connected with an effort to get a pension. It is possible that a search of the old Adjutant General Records at State Archives may turn up the original letter. However, there is another way of reading this note. During the Civil War, there was an army hospital named "Mount Pleasant" in Washington D.C. It is possible that "09" is really "64" (bad handwriting), and the correct date of the letter is 8/18/64.

 All Rights Reserved by Chris Czopek

SAMUEL C. CHATFIELD 023

Chatfield, W. Samuel, Isabella. Enlisted in company K, First Sharpshooters, May 18, 1863, at Isabella, for 3 years, age 26. Mustered May 26, 1863. Killed in action at Weldon R. R., Va., Aug. 19, 1864.

Chatfield, Samuel C. – PRIME NAME, from Pension File
Chatfield, W. Samuel – Adjutant General Record name
Chalfield, Samuel C. – Pension File Index Card, Organization Card

RESIDENCE IN 1863: Isabella, Isabella County S-AGR

DEATH DATE: August 19, 1864 (*Killed in action near Petersburg, VA*)

GRAVE: Location Unknown

PENSION FILE: Yes, see Appendix A
1883 Pension Roll: No

VETERAN CENSUS: 1888 No / 1890 No / 1894 No

RAY HEREK's BOOK: Pages: 243, 404, 434(Roster), 488, 489

SPECIAL INFORMATION:
Official army records (Adjutant General Record) give his name as "W. Samuel Chatfield", but his pension file says "Samuel C. Chatfield". I say the pension file name is correct.
Samuel C. Chatfield, age 26, enlisted on the same day & same place as Charles Chatfield, age 19. They were probably related (cousins?) and decided to join the army together.
Samuel Chatfield was killed in action near Petersburg during a battle at Weldon Railroad.

NOTE TO HISTORIANS:
The fighting at Weldon Railroad was near the historic Globe Tavern, which locals called "The Yellow House". S-pf-023, and S-Herek pg 241
The Globe Tavern was torn down after the war, and the railroad tracks were removed.
Nothing remains today at the battle site but a vacant field and a Historical Marker.

 All Rights Reserved by Chris Czopek

ALBERT CHURCH 024

Morse, or Moses, James, Venice. Enlisted in company M, Eighth Cavalry, April 28, 1863, at Venice, for 3 years, age 21. Mustered May 2, 1863. Deserted at Mt. Clemens, Mich., May 22, 1863.

Church, Albert, St. Charles. Enlisted in company K, First Sharpshooters, June 10, 1863, at East Saginaw, for 3 years, age 20. Mustered June 16, 1863. Mustered out at Delaney House, D. C., June 28, 1865.

Church, Albert – PRIME NAME, Adjutant General Record name
Church, James – Gravestone name
Chirch, Albert – Pension File name (notice the letter "i")
Morse / Moses, James – Name used for enlistment in the 8th Cavalry

RESIDENCE IN 1863: St. Charles, Saginaw County S-AGR
DEATH DATE: February 1875 S-pf-024
GRAVE: Saginaw County, St. Charles / Riverside Cemetery (Military stone)
Buried in the east half of the cemetery, Section B, Row 1, name on gravestone is "James Church"
PENSION FILE: Yes, see Appendix A (filed under name "Chirch, Albert")
1883 Pension Roll: No
VETERAN CENSUS: 1888 No / 1890 No / 1894 No
RAY HEREK's BOOK: Pages: 434(Roster)

SPECIAL INFORMATION:
In the woods near St. Charles, Michigan, there was once a thriving Chippewa settlement the local people called "Indiantown". During the Civil War, some of the young men living there joined the army to fight for the Union. Albert Church was one of those volunteers. For his first enlistment he went to Venice township (in Shiawassee County) and joined the 8th Michigan Cavalry under the name "James Moses" S-pf-024. Three weeks later he went AWOL and enlisted in Company K of the 1st Michigan Sharpshooters. To the government, deserting his cavalry unit was an unforgivable offense. To modern historians, it is understandable that when Albert found out about an all-Indian unit of Sharpshooters, he would risk severe punishment to be with "his people". Whatever the reason, he enlisted in Company K using his real name, Albert Church. He fought on the battlefields of Virginia and was twice wounded (at the Wilderness and Petersburg). Perhaps his closest call with death was a case of Typhoid Fever. He survived the war, received an honorable discharge, and returned to Indiantown. But he never recovered his health; Albert Church died in 1875 from chronic diarrhea. When his widow applied for a pension, her claim was denied. The government never forgave Albert for deserting from the 8th Michigan Cavalry. Consequently, the wife and two young daughters of this battle-scarred veteran were left to fend for themselves.

NOTES FOR HISTORIANS:
Error in official records: Albert Church was mustered out on July 28, not June 28. S-pf-024
Family Information: Wife = Julia David, b.1848. There were two daughters: Nancy b.1870, and Susan b.1874. Julia was helped by Peter David, b.1847, who was probably her brother or cousin.

JACOB COLLINS 025

Collins, Jacob, Isabella. Enlisted in company K, First Sharpshooters, June 10, 1863, at East Saginaw, for 3 years, age 30. Mustered June 16, 1863. Absent wounded. No discharge given.

Collins, Jacob – PRIME NAME, all sources agree on spelling

RESIDENCE IN 1863: Isabella, Isabella County S-AGR

DEATH DATE: Winter of 1867-1868 S-Shier page 98

GRAVE: Location Unknown (*he died in Isabella County*)

PENSION FILE: Yes, see Appendix A
1883 Pension Roll: No

VETERAN CENSUS: 1888 No / 1890 No / 1894 No

RAY HEREK's BOOK: Pages: 226, 234, 402, 434(Roster)

SPECIAL INFORMATION:
There were three soldiers in Company K named Collins: Jacob, John, and William. S-AGR They were all from Isabella. Were these three men related? Future research needed.
Jacob Collins was badly wounded at the Battle of the Crater at Petersburg, VA, on July 30, 1864. His left arm was smashed by a bullet (S-Herek pg 226 and 234). After the fight he was sent away to a hospital, and his regiment lost track of him. When Company K was mustered out of service after the end of the war, his fate was unknown. A note was made in his records "No discharge given" and the books were closed.
Today we know what happened to Jacob. He did receive a discharge at the end of the war and was mustered out of service at a hospital in Washington D.C. on April 13, 1865. He returned home to Isabella County and married a woman named Eliza. But Jacob was not a well man. He applied to the government for a disability pension. Unfortunately, he died before it was approved. His death occurred in the winter of 1867-1868. S-Shier page 98

JOHN COLLINS 026

Collins, John, Isabella. Enlisted in company K, First Sharpshooters, May 18, 1863, at Isabella, for 3 years, age 21. Mustered May 26, 1863. Mustered out at Delaney House, D. C., July 28, 1865.

Collins, John – PRIME NAME, all sources agree on spelling

RESIDENCE IN 1863: Isabella, Isabella County S-AGR

DEATH DATE: Unknown

GRAVE: Location Unknown

PENSION FILE: No Pension File exists

VETERAN CENSUS: 1888 No / 1890 No / 1894 No

RAY HEREK's BOOK: Pages: 434(Roster)

SPECIAL INFORMATION:
There were three soldiers in Company K named Collins: Jacob, John and William.
All of them were from Isabella.
John and William enlisted together on the same day at the same place. Were they related? Future research needed.
John Collins served for a while as a Hospital Attendant, helping take care of the many sick and wounded (S-Descriptive Roll, State Archives). He was discharged at the end of the war with the rest of the regiment - then disappears. Nobody knows where he lived after the war, or when he died.

WILLIAM COLLINS 027

Collins, William, Isabella. Enlisted in company K, First Sharpshooters, as Corporal, May 18, 1863, at Isabella, for 3 years, age 33. Mustered May 26, 1863. Deserted at Chicago, Ill., Oct. 2, 1863.

Collins, William – PRIME NAME, all sources agree on spelling

RESIDENCE IN 1863: Isabella, Isabella County S-AGR

DEATH DATE: February 7, 1899 S-State Death Record

GRAVE: Bay County, Bay City / Oak Ridge Cemetery, in the "Indian Corner"
S- West Bay City Times-Press, February 9, 1899, Page 1, Column 1

PENSION FILE: No Pension File exists

VETERAN CENSUS: 1888 No / 1890 No / 1894 No

RAY HEREK's BOOK: Pages: 34, 434(Roster)

SPECIAL INFORMATION:
There were three soldiers in Company K named Collins: Jacob, John and William.
They all came from Isabella. John and William enlisted together on the same day at the same place. Were they related? Future research needed.
William Collins was an interpreter. After enlisting in Company K, he helped the officers recruit men from the Chippewa tribe for Company K. S-Herek pg 34
The first assignment for the 1st Michigan Sharpshooters was guarding Confederate prisoners in Chicago. Most soldiers in the regiment had joined to the army to fight in battles. Guard duty was a bitter disappointment. A number of men deserted. William Collins was one.
What happened to him after he left the army is a mystery. One record says that in 1899 he was living in Cheboyganing, Saginaw County. We know this because he gave testimony for the pension of a fellow soldier, William Cabecoung. S-pf-019 (pension file for William Cabecoung)
William Collins died February 7, 1899 in Bangor township of Bay County. S-State Death Record
His death was front page news in the Bay City newspapers. According to one newspaper, he was buried in the "Indian Corner" of Bay City's Oak Ridge Cemetery. S- West Bay City Times-Press, February 9, 1899, Page 1, Column 1

GEORGE CORBIN 028

Corbin, George, Isabella. Enlisted in company K, First Sharpshooters, as Corporal, May 18, 1863, at Isabella, for 3 years, age 23. Mustered May 26, 1863. Promoted Sergeant May 1, 1865. Mustered out at Delaney House, D. C., July 28, 1865.

Corbin, George – PRIME NAME, all sources agree on spelling

RESIDENCE IN 1863: Isabella, Isabella County S-AGR
In 1888: Delwin, Denver township, Isabella County S-pf-059 (Kabayacega pension file)
In 1894: Denver township, Isabella County S-1894 Veteran Census
In 1928: Died at home in Denver township, Isabella County S-pf-028

DEATH DATE: May 20, 1928 S-pf-028
(*Died at home in Denver township, Isabella County*)

GRAVE: Location Unknown (*A descendant told me Amos owned a plot in River Lawn Cemetery in Denver township, Isabella County, and he thought that's where Amos is buried*)

PENSION FILE: Yes, see Appendix A
1883 Pension Roll: No

VETERAN CENSUS: 1888 No / 1890 No / 1894 Yes

RAY HEREK's BOOK: Pages: 434(Roster)

SPECIAL INFORMATION:
When George Corbin died in 1928 at age 97, he was one of the last surviving members of Company K. Only one man lived longer, Moses Thomas.
His death was front page news in the local paper. S-Newspaper, Mt. Pleasant Daily Times, May 21, 1928, Page 1, Col 2.
In 1886, George Corbin joined the G.A.R., a veterans organization like the VFW of today. He was a member of the Wa-Bu-No G.A.R. Post No. 250, Mt. Pleasant, Isabella County.
During the war, George Corbin was promoted to Sergeant.

NOTE TO HISTORIANS:
It was the custom to name a G.A.R. Post after a local soldier who had died in the war.
Mt. Pleasant's Post No. 250 was named after a soldier in Company K, Thomas Wabano, who died of disease while home on leave. Of the several thousand G.A.R. Posts across the country, this was the only G.A.R. Post named after a Native American soldier.

AMOS CRANE 029

Crane, Amos. Enlisted in company K, First Sharpshooters, Aug. 24, 1864, at Jackson, for 3 years, age 28. Mustered Aug. 24, 1864. Substitute for DeWitt C. Leach. Discharged at Detroit, Mich., Aug. 11, 1865.

Crane, Amos – PRIME NAME, all sources agree on spelling

RESIDENCE IN 1863: Living in the Upper Peninsula S-Herek pg 250
Credited to Lansing for draft quota purposes S-Descriptive Roll, State Archives
In 1890: Baraga County (in the U.P.), by the historic Kewawenon Mission (aka "the Methodist Mission") at Zeba, near L'Anse S-1890 Veteran Census

DEATH DATE: April 14, 1891 S-Newspaper: L'Anse Sentinel, April 18, 1891, Pg 5, Col.1

GRAVE: Upper Peninsula, Baraga County, L'Anse / Pinery Cemetery S-ccn029a
(There is no marker on his grave, and no record of where inside the cemetery he is buried. His death record says he was buried in the Methodist Mission Cemetery, which is another name for the Pinery Indian Cemetery)

PENSION FILE: Yes, see Appendix A; 1883 Pension Roll: No

VETERAN CENSUS: 1888 No / 1890 Yes / 1894 No

RAY HEREK's BOOK: Pages: 250, 434(Roster)

Amos Crane came from Michigan's Upper Peninsula. He was probably a member of the historic L'Anse band of the Lake Superior Chippewa, which is known today as the Keweenaw Bay Indian Community. He had strong ties with the famous Kewawenon Mission, a very old Methodist Mission at Zeba near L'Anse, in Baraga County. L'Anse Sentinel, April 18, 1891, Pg 5, Col.1
The records of his enlistment into the army have many surprises. First he was recruited in the Upper Peninsula (S-Herek pg 250). Then he officially enlisted in Jackson, but was credited to Ingham County for draft quota purposes. Next, he signed papers saying he entered the army as a substitute for DeWitt C. Leach. This was no ordinary man; DeWitt C. Leach was the Indian Commissioner for the State of Michigan, personally appointed by Abraham Lincoln. When the draft started, there was concern that high ranking politicians would be exempt. Abraham Lincoln countered this by paying a man to be his substitute for army service! It appears that DeWitt Leach was following Lincoln's example. And since he was the Indian Commissioner, he paid an Indian to be his substitute.
After the war, Amos Crane returned to the UP and lived near L'Anse at Zeba as part of the Kewawenon Mission community. The 1880 Census records show that he was a farmer with a wife named "Sarah" and three young children. The 1890 Veteran Census shows he received his mail at the "Methodist Mission". He died in 1891. His widow "Charlotte" applied for a pension. S-pf-029
In reporting his death, the local newspaper called him an "old and respected" man and mentioned he was a veteran of the Civil War. S-Newspaper: L'Anse Sentinel, April 18, 1891, Pg 5, Col 1

JONAH DABASEQUAM 030

Dabasequam, Jonah, Little Traverse. Enlisted in company K, First Sharpshooters, June 24, 1863, at Little Traverse, for 3 years, age 20. Mustered July 20, 1863. Killed in action at Spottsylvania, Va., May 12. 1864.

Dabasequam, Jonah – PRIME NAME, Adjutant General Record
Labasquam / Labosquam, Jona – Pension File name
Dabusgum, Jonah – Adjutant General Annual Report

RESIDENCE IN 1863: Little Traverse (=Harbor Springs), Emmet County S-AGR
Childhood: It appears he was born & raised on the Old Mission Peninsula in Grand Traverse Co. S-pf-030

DEATH DATE: May 12, 1864 (*Killed in action at the Battle of Spotsylvania, VA*) S-AGR

GRAVE: Location Unknown

PENSION FILE: Yes, see Appendix A (filed under name "Labosquam, Jona")
1883 Pension Roll: No

VETERAN CENSUS: 1888 No / 1890 No / 1894 No

RAY HEREK's BOOK: Pages: 351, 389, 434(Roster)

SPECIAL INFORMATION:
Killed in action at the Battle of Spotsylvania.
The correct spelling of his name is a problem: Neither Dabasequam nor Labasquam sound right to experts on the Anishinabe language. Some say the D should be a T, and his correct name is Tabasquam. Others say the L should be the letter I, and the correct name is Iabasquam. Further research needed.

Here is what we know about his life: Jonah was born and raised on Old Mission Point in Grand Traverse Bay. He joined the army when he was 20. Killed by an exploding shell at the Battle of Spotsylvania and buried on the battlefield in an unmarked grave. He left behind neither wife nor children, just an old father in Antrim County near Kewadin. The father, John Labasquam, applied for a pension in 1886, saying his son was his only means of support in old age. The paperwork took years to complete. Just as the pension was about to be approved, the father passed away.
S-pf-030

JOHN DAVID 031

David, John, Isabella. Enlisted in company K, First Sharpshooters, May 19, 1863, at East Saginaw, for 3 years, age 21. Mustered May 26, 1863. Died of disease at Annapolis, Md., May 4, 1864. Buried at Annapolis, Md.

David, John – PRIME NAME, Adjutant General Record
Davis, Jno – Gravestone and Cemetery Records
Ta-ba-se-ka / To-ba-se-ka - Name on Enlistment Papers

RESIDENCE IN 1863: Isabella, Isabella County. S-AGR

DEATH DATE: May 4, 1864 (*Died of disease in an army hospital*)

GRAVE: Annapolis National Cemetery, 800 West Street, Annapolis, MD., Section K, Grave # 1078
*His grave is near the flagpole, and the name on his gravestone is "**Jno. Davis**".*

PENSION FILE: Yes, see Appendix A; 1883 Pension Roll: No

VETERAN CENSUS: 1888 No / 1890 No / 1894 No

RAY HEREK's BOOK: Pages: 416, 434(Roster)

SPECIAL INFORMATION:
He was a Chippewa from the Saginaw area. On his enlistment paper, his name was entered as "John David, in Indian Ta-ba-se-ka". His Indian name also spelled To-ba-se-ka & Tay-Bay-Se-Gay. His military career was short - died from disease before he saw his first battle. Specifically, he died of typhoid fever at a military hospital in Annapolis, Maryland.

Notes for locating his grave: Annapolis National Cemetery is a small cemetery in downtown Annapolis, surrounded by hotels and banks. It is many miles away from the famous Naval Academy. Address: 800 West Street. His grave is near the cemetery flagpole, stone #1078. **Name on his gravestone is "Jno. Davis". Cemetery records have him as "John Davis".**

NOTES FOR HISTORIANS:
This man is named John David. Be careful not to confuse him with another soldier named David Johns. Both men were in Company K; it is easy to mistake one for the other. John David was not married. On the 1890 Veteran Census, the widow living in Mason County named Mary Kawis is the widow of David Johns.

JAMES S. DELAND 032

DeLand, James S., Jackson. Enlisted in company F, First Sharpshooters, Dec. 29, 1863, at Jackson, for 3 years, age 28. Mustered Jan. 19, 1864. Promoted Sergeant Major April 15, 1864. Promoted First Lieutenant, company K, June 20, 1864. Mustered Sept. 20, 1864. Promoted Captain Nov. 20, 1864. Mustered Dec. 14, 1864. Wounded in action July 11, 1864, and April 2, 1865, at Petersburg, Va. Brevet Major, U. S. Volunteers, for gallant and meritorious services in the assault before Petersburg, Va., April 2, 1865. Mustered out at Delaney House, D. C., July 28, 1865. Present residence, Napoleon, Mich.

DeLand, James S. – PRIME NAME, all sources agree on spelling

RESIDENCE IN 1863: Jackson, Jackson County S-AGR
In 1890: Napoleon, Jackson County S-1890 Veteran Census

DEATH DATE: October 29, 1925 S-pf-032 *(Died in Detroit, Wayne County)*

GRAVE: Jackson County, Jackson / Mount Evergreen Cemetery (Civilian stone)
His grave can been found in the North West corner of cemetery, in the middle of Section W, Block 8, Range 2, Lot 5 See FIND A GRAVE MEMORIAL # 16354426

PENSION FILE: Yes, see Appendix A; 1883 Pension Roll: No

VETERAN CENSUS: 1888 No / 1890 Yes / 1894 Yes

RAY HEREK's BOOK: Pages: 88, 95, 137, 138, 150, 195, 197, 210, 236, 248, 251, 262, 263, 281, 283, 297, 304, 306, 307, 317, 319, 320, 344, 346, 353, 377, 382, 401, 407, 428(Roster), 497

SPECIAL INFORMATION:
White officer (1st Lieutenant and Captain) in Company K.
He was the younger brother of the regiment commander, Colonel Charles V. DeLand.
He started out as an enlisted soldier in Company F. Promoted to 1st Lieutenant of Co. K. When Captain Andress was discharged on account of his wounds, Lt. James DeLand was promoted to Captain to take his place. **The soldiers of Company K felt honored the Commander chose his own brother to be their leader on the battlefield.** S-ccn032a
Promoted to the rank of Major for gallant conduct while leading Company K into battle on April 2, 1865. During this attack he was severely wounded and lost the use of his left arm.

NOTE TO HISTORIANS: Correct way to spell his last name is DeLand, with a capitol "L". Captain DeLand lived till 1925. Did anyone ever write down his stories about the war?

WILLIAM J. DRIGGS 033

Driggs, William J. Enlisted in company L, Sixth Cavalry, as Corporal, Aug. 29, 1862, at East Saginaw, for 3 years, age 23. Mustered Oct. 11, 1862. Transferred to company C, Seventh Cavalry, Nov. 12, 1862. Commissioned First Lieutenant, company K, First Sharpshooters, at organization, July 22, 1863. Mustered Dec. 20, 1863. Discharged for disability July 6, 1864. Second Lieutenant, Eleventh U. S. Infantry, Feb. 23, 1866. First Lieutenant July 28, 1866. Transferred to Twentieth Infantry Sept. 21, 1866. Mustered out Jan. 1, 1871. Present residence, Saginaw, Mich.

Driggs, William J. – PRIME NAME, most records agree on spelling
Driggs Jr., William – Forest Lawn Cemetery records, Saginaw

RESIDENCE IN 1863: East Saginaw, Saginaw County S-Descriptive Roll, State Archives

DEATH DATE: December 13, 1914 (*Died in Evart, Osceola County*)

GRAVE: Saginaw County, Saginaw / Forest Lawn Cemetery, Section 8, Lot 189
He has no gravestone. Listed on cemetery records as "William Driggs Jr."

PENSION FILE: Yes, see Appendix A; 1883 Pension Roll: No

VETERAN CENSUS: 1888 No / 1890 No / 1894 No

RAY HEREK's BOOK: Pages: 34, 35, 36, 155, 190, 374, 434(Roster)

SPECIAL INFORMATION:
White officer (1st Lieutenant) in Company K.
He was the original 1st Lieutenant of Company K. As in the Colored troops, all officers of Company K were white men (except 2nd Lieutenant Garrett Graveraet).
William was the son of a prominent Republican congressman from Saginaw, John F. Driggs.
When chief Daniel Mwakewenah of Company K died from wounds in a Washington hospital, Lt. Driggs and his father, Congressman Driggs, made arrangements to have the body embalmed, sent back to Michigan, and buried in Brady Hill cemetery, Saginaw.
Lt. Driggs had suffered from poor health before he joined the Sharpshooters. S-Herek pg 34
He lasted for a few battles, and then was discharged for disability in July of 1864.
Driggs belonged to Gordon Granger G.A.R. Post No. 38, East Saginaw, Saginaw County.
He was also a member of the Sedgwick G.A.R. Post No. 16, Evart, Osceola County.
State Death Records say he died on Dec 13, 1914, but his newspaper obituary says Dec 12, 1914.
His grave has no gravestone! The family plot has one large stone with seven names on it, names of the Driggs family members buried there in unmarked graves. His name is third from the top.

 All Rights Reserved by Chris Czopek

LUKE DUTTON 034

Dutton, Luther, Taymouth. Enlisted in company K, First Sharpshooters, June 2, 1863, at East Saginaw, for 3 years, age 19. Mustered June 16, 1863. Wounded in action June 17, 1864. Discharged at Detroit, Mich., May 12, 1865.

Dutton, Luke – PRIME NAME, from Pension File
Dutton, Luther – Adjutant General Record name

RESIDENCE IN 1863: Taymouth township, Saginaw County S-AGR

DEATH DATE: June 1870 S-carved on his gravestone

GRAVE: Saginaw County, Taymouth township / Taymouth Indian Cemetery
(This cemetery is also called Wheaton Indian cemetery)
Name on gravestone is "Luke Dutton"

PENSION FILE: Yes, see Appendix A
1883 Pension Roll: No

VETERAN CENSUS: 1888 No / 1890 No / 1894 No

RAY HEREK's BOOK: Pages: 396, 434(Roster)

SPECIAL INFORMATION:
Official army records (The Adjutant General Record) say his first name was Luther; but his pension file has testimony from the family giving his correct name as Luke Dutton.
In fact, the testimony in the pension file states that his family and friends were puzzled by the name Luther. They had no idea why the army called him "Luther". Maybe a clerical error?
His gravestone has the correct name, Luke Dutton.
Wounded in the leg on June 17, 1864, at Petersburg, VA.
Medical records say he received a gunshot wound in the right thigh.
Luke Dutton died of Consumption at his father's house in Taymouth in June, 1870.
This information comes from his pension file. No specific date in June is given for his death.
Consumption was the old fashioned name for tuberculosis, a lung disease.

JOHN ETARWEGESHIG 035

E-tar-ivegezhik, John, Pentwater. Enlisted in company K, First Sharpshooters, July 5, 1863, at Pentwater, for 3 years, age 26. Mustered July 11, 1863. Killed in action at Spottsylvania, Va., May 12, 1864.

Etarwegeshig, John – PRIME NAME, from Pension File
E-tar-ivegezhik, John – Adjutant General Record name
E-tar-we-geeshig, Tah-te-go-quah – 1883 Pension Roll (mother)
Waw-me-gwan, John – Name he used in letters to his mother

RESIDENCE IN 1863: Pentwater, Oceana County S-AGR

DEATH DATE: May 12, 1864 (*Killed in action at Spotsylvania, VA*)

GRAVE: Location Unknown

PENSION FILE: Yes, see Appendix A
1883 Pension Roll: Yes - Oceana Co. – Hart (Mother "Tah-te-go-quah")

VETERAN CENSUS: 1888 No / 1890 No / 1894 No

RAY HEREK's BOOK: Pages: 389, 434(Roster)

SPECIAL INFORMATION:
Adjutant General Record name is E-tar-ivegezhik, an obvious clerical error. Change the "iv" to the letter "w" and the name will sound right. Since his pension file spells his name as "Etarwegeshig", my decision is to use this spelling for his Prime Name.
Killed in action at Spotsylvania, VA, on May 12, 1864.
There is a Veterans Monument in Hart, Michigan (Oceana County) with his name on it.

The mother of John Etarwegeshig was named Pah-te-go-quay / Po-te-go-quah and lived near Hart in Oceana County. She applied to the government for a pension in 1869. Her claim was based on the fact that she had depended on her son for support. To prove her case, she sent to the pension department two letters written to her by her son. **These are among the few existing letters written home by a Company K soldier during the war**. Unfortunately, there is strong evidence indicating the government delayed granting her pension for no other reason than she was an Indian. Finally, the US Indian Agent for Michigan in the 1870's, George W. Lee, took up the case. In a letter dated June 22, 1877, he bluntly says "It really seems that if this case had been a white woman, or a white soldier, these queries raised all these years would not be thought of ". The government finally accepted the claim and paid her $8.00 a month for the rest of her life. When she died July 18, 1902, they wrote on her death certificate "age estimated over 100 years".

 All Rights Reserved by Chris Czopek

LOUIS GENEREAU Jr. 036

Genereau, Louis, Pentwater. Enlisted in company K, First Sharpshooters, as Sergeant, July 4, 1863, at Elbridge, for 3 years, age 18. Mustered July 11, 1863. Mustered out at Delaney House, D. C., July 28, 1865.

Genereau Jr., Louis – PRIME NAME S-State Death Record(GENDIS) & Gravestone
Genereau, Louis – Adjutant General Record name
Genereaux, Louis – aka name in 1890 Veteran Census / Index book

RESIDENCE IN 1863: Pentwater, Oceana County S-AGR
In 1871: Elbridge township, Oceana County S-State Death Record(GENDIS)

DEATH DATE: June 25, 1871
(Died of Consumption in Elbridge township, Oceana Co.) S-State Death Record(GENDIS)

GRAVE: Oceana Co., Elbridge twp. / Methodist Indian Cemetery on Tyler Road (on the north side of Tyler Road, near the intersection with 120th Street) S-ccn036a
The cemetery is also called Genereau Family Burying Ground

PENSION FILE: Yes, see Appendix A; 1883 Pension Roll: No
VETERAN CENSUS: 1888 No / 1890 Yes(widow) / 1894 No
RAY HEREK's BOOK: Pages: 35, 148, 186, 390, 434(Roster)

SPECIAL INFORMATION:
Most records refer to him as "Louis Genereau", but his correct name was "Louis Genereau Jr.".
Genereau is the standard spelling, but a few records say "Genereaux".
His gravestone says "Louis Genereau Jun.", Jun. = Junior.
His father, Louis Genereau, was a well-known and influential Interpreter for the tribes living in the Pentwater-Elbridge area of Oceana County.
Louis Genereau Jr's place of birth is Grand River. S-State Death Record(GENDIS)
The location of his birthplace is not clear. "Grand River" was the original name for Delta Mills, an early settlement in Eaton County on the Grand River. Perhaps that is where he was born?
Wounded at Spotsylvania, VA – shot through his left leg, near the ankle. S-pf-036
While home on leave to recover from his wound, Louis Genereau fell in love and was married on August 19, 1864, to Louisa Duvernay/Deverney. She was the sister of John Kedgnal (Co. B, 1st Mich SS). They had one child who survived to adulthood, a son named Levi Genereau. S-pf-036
Sergeant Genereau was captured in battle, but managed to escape! S-Herek pg 186
"The Tree That Never Dies - Oral History of the Michigan Indians" by Pamela Dobson, 1978, has a story about Louis Genereau Jr. in the war. This story was passed down by word of mouth for more than one hundred years among Odawa families.
The 1890 Veteran Census shows his widow living in Petoskey, Emmet Co., as "Louisa Keway".
She died April 29, 1918, and is buried in Greenwood Cemetery, Petoskey, as Louisa Genereau.

DAVID GEORGE 037

George, David, Isabella. Enlisted in company K, First Sharpshooters, May 18, 1863, at Isabella, for 3 years, age 22. Mustered May 26, 1863. Died May 26, 1864, of wounds received in action at Spottsylvania, Va., May 12, 1864. Buried in National Cemetery at Arlington, Va.

George, David – PRIME NAME, all sources agree with spelling

RESIDENCE IN 1863: Isabella, Isabella County S-AGR

DEATH DATE: May 26, 1864 (*Died in a Washington D.C. hospital from battle wounds*)

GRAVE: Arlington National Cemetery, Washington D.C., Section 27, # 325

PENSION FILE: Yes, see Appendix A; 1883 Pension Roll: No

VETERAN CENSUS: 1888 No / 1890 No / 1894 No

RAY HEREK's BOOK: Pages: 390, 434(Roster)

SPECIAL INFORMATION:
Buried in Arlington National Cemetery, Washington D.C.
The following description of his life is based on information from his pension file -
David George was a Chippewa living on the Isabella reservation.
His father was named George Nagatoe. When David joined the army, he used his father's first name as his last name, and became David George.
He sent money to his parents on the Isabella reservation every time he was paid.
Wounded while making a charge during the battle of Spotsylvania, VA, on May 12, 1864.
Arrived at Emery Hospital, Washington D.C., on May 25, 1864. Died the following day.
He was never married and left behind neither wife nor children.
His mother, Nancy Nagatoe, received a pension from the government for the rest of her life.

NOTES FOR HISTORIANS:
David George had been wounded in the head. Exact details of his injury can be found here:
Medical/Surgical History--Part I, Volume II, Chapter I.--Wounds And Injuries Of The Head. Section III--Gunshot Contusions Of The Cranial Bones.

Caution: State Archives of Michigan has a list titled "Report of Michigan Soldiers who have died in Hospitals, 1862-1865". This list says David George died in Emory hospital on May 27, of wounds. The date is wrong, George died May 26. S-pf-037 (pension file for David George).

JOSEPH L. GIBSON 038

Gibson, Joseph, or Louis G., Bear Creek. Enlisted in company K, First Sharpshooters, Sept. 8, 1863, at Little Traverse, for 3 years, age 23. Mustered Oct. 21, 1863. Taken prisoner at Petersburg, Va., June 17, 1864. Died in prison at Andersonville, Ga., Sept. 3, 1864. Buried in National Cemetery at Andersonville, Ga. Grave No. 7741.

Gibson, Joseph L. – PRIME NAME, Adjutant Gen. Annual Report
Gibson, J. L. – Name on his gravestone at Andersonville
Gibson, Louis J. – His signature on a letter
Gibson, Louis G. – aka name in Adjutant General Record
Nabawnayasang, Joseph – aka name, Pension File

RESIDENCE IN 1863: Bear Creek township, Emmet County S-AGR

DEATH DATE: September 3, 1864 (*Died while a POW at Andersonville Prison*) S-AGR

GRAVE: Georgia / Andersonville National Cemetery, Grave No. 7741
GPS N32.20448 W84.13179

PENSION FILE: Yes, see Appendix A; 1883 Pension Roll: No

VETERAN CENSUS: 1888 No / 1890 No / 1894 No

RAY HEREK's BOOK: Pages: 288, 398, 413, 434(Roster)

SPECIAL INFORMATION:
In Traverse City there is a Veterans Park at 11th Street and Elmwood Avenue. In this park is a block of stone with the names of local men who died in the Civil War. "Joseph Gibson" is one of the names on this memorial.
There is a problem with his first name. Most records give his name as either Joseph Gibson or Joseph L. Gibson. However, his pension file contains a letter he wrote, and he signed his name "Louis J. Gibson". And the Adjutant General Record gives "Louis G. Gibson" as an aka name. Since the most important records say Joseph was his first name, I choose it as his Prime Name. According to testimony in his pension file, his Anishinabe name was Joseph Nabawnayasang. One story says the name "Gibson" was assigned to him at random by a school teacher. S-pf-038
Genealogist Quita V. Shier claims Joe deliberately chose the name Gibson because it was the name of a teacher that he admired, a teacher named Catherine Gibson. S-Shier pg 132
He was a resident of Bear Creek township, not Bear River township. Both places are in Emmet Co.
Joseph L. Gibson was one of 15 soldiers in Company K captured on June 17, 1864 (same battle where Lt. Graveraet was mortally wounded) and sent to the infamous Andersonville prison. He was the first of the group to die there from starvation and disease.

SAMUEL GOING 039

Going, Samuel, Pentwater. Enlisted in company K, First Sharpshooters, July 4, 1863, at Elbridge, for 3 years, age 20. Mustered July 11, 1863. Corporal. Killed in action at Spottsylvania, Va., May 12, 1864.

Going, Samuel – PRIME NAME, Adjutant General Record
Goeng, Samuel – Pension File name
Gowing, Samuel – Adjutant General Annual Report
Gora, Samuel – Name on gravestone and cemetery records S-ccn039a

RESIDENCE IN 1863: Pentwater, Oceana County S-AGR

DEATH DATE: May 10, 1864 S-pf-039 (*Mortally wounded May 9 at Ni River, VA*)

GRAVE: Fredericksburg National Cemetery, VA., Grave No. 2186
(GPS 38.29305, -77.4689) Buried as "Samuel Gora" S-ccn039a

PENSION FILE: Yes, see Appendix A
1883 Pension Roll: No

VETERAN CENSUS: 1888 No / 1890 No / 1894 No

RAY HEREK's BOOK: Pages: 148, 389, 434(Roster)

SPECIAL INFORMATION:
There is much conflicting information about the exact date and place of his death in 1864. Official Army Records say he was killed in action at Spotsylvania on May 12. S-AGR However, documents in his pension file say he received a gunshot wound to his head on May 9, which would be Battle of Ni River, and died in an army hospital the following day, May 10. Fredericksburg National Cemetery has a grave for "Samuel Gora, Co. K, 1 Mich", and their records say he died on May 20 or May 23 (different sources give different dates) at the hospital at Marye's Heights, Fredericksburg, VA.
There is a Veterans Monument in Hart, Michigan (Oceana County) with his name on it.

NOTES FOR HISTORIANS:
Detroit Advertiser Tribune, July 22, 1864, front page, says "Samuel Gora (or Gudian) of the 1st Michigan Sharpshooters died in hospital at Fredericksburg". I believe "or Gudian" = "an Indian". In his pension file, Sam's mother says "Tondoqua" is the Indian word for Going.

GARRETT A. GRAVERAET 040

Gaveraet, Garrett A., Little Traverse. Entered service in company K, First Sharpshooters, at organization, as Second Lieutenant, June 9, 1863, at Little Traverse, for 3 years, age 23. Commissioned July 22, 1863. Mustered July 23, 1863. Died July 10, 1864, at Washington, D. C., from wounds received in action before Petersburg, Va., June 17, 1864. Buried in family lot on Mackinac Island.

Graveraet, Garrett A. – PRIME NAME S-Descriptive Roll, State Archives
Gaveraet, Garrett A. – Adjutant General Record name (misspelled)
Graveraet, Gerrett A. – Pension File name

RESIDENCE: He grew up on Mackinac Island S-1850 US Census, and is buried there.
In 1863: Little Traverse (=Harbor Springs), Emmet County S-AGR
DEATH DATE: June 30, 1864 (Not July 10, 1864) S-pf-040
(*Wounded at Petersburg, VA, on June 17. Died in an army hospital in Washington D.C.*)
GRAVE: Mackinac Co., Mackinac Island / St. Anne Cemetery (Military stone)
PENSION FILE: Yes, see Appendix A
1883 Pension Roll: Yes, Emmet Co. – Harbor Springs (mother, Sophia Groouat)
VETERAN CENSUS: 1888 No / 1890 No / 1894 No
RAY HEREK's BOOK: Pages: 35, 36, 148, 150, 179, 182, 344, 351, 395, 434(Roster), 450

SPECIAL INFORMATION:
He was 2nd Lieutenant of Company K and a Native American.
During the siege of Petersburg, he was ordered to lead Company K in a daring attack on a rebel battery. The attack succeeded, but the gallant Lieutenant Graveraet was wounded and sent to a hospital in Washington D.C. There, his left arm was amputated and he died from complications. Official army records (Adjutant General Record) say he died on July 10, which is wrong. His gravestone says he died July 1, also wrong. **Correct death date is June 30, 1864.** S-pf-040 (hospital records in his pension file). After the war, his body was sent home to be buried on Mackinac Island.
His name appears on three monuments : The monument to the 1st Michigan Sharpshooters on the front lawn of the State Capitol building in Lansing, the new veterans monument on Mackinac Island, and the Traverse City Veterans Park Monument (11th Street and Elmwood Avenue), where a block of stone listing Civil War dead has the name "G A GRAVERAT".
See page 220 of this book for his obituary in a Detroit newspaper – fascinating details of his life.
Official army records (Adjutant General Record) spell his name "Gaveraet" which is wrong.
His pension file index card spells his first name Gerrett; all other sources spell it Garrett.
His father, Henry G. Graveraet Jr., also enlisted in Company K. His rank was First Sergeant.
Garrett Graveraet was an educated and refined young man. Before he became a soldier, he was a teacher at a government school for Native Americans in Harbor Springs. S-ccn040a
In Marquette, Michigan, Graveraet High School was named after his aunt Juliet. S-ccn040b
The family name, Graveraet, is a Dutch name. Garrett's ancestors came from Holland and settled in New York near Albany. They migrated from Albany to Detroit around 1772. S-Shier pg 138

HENRY G. GRAVERAET Jr. 041

Graveraet, Henry G., Little Traverse. Enlisted in company K, First Sharpshooters, as First Sergeant, July 7, 1863, at Dearborn, for 3 years, age 45. Mustered July 20, 1863. Killed in action at Spottsylvania, Va., May 12, 1864.

Graveraet Jr., Henry G. – PRIME NAME, Genealogy Records
Graverat, H. Garrat – Name on the monument in Traverse City
Graverates, Henry – Name on gravestone (Fredericksburg, VA)
Mankewenan – Said to be his name (see note below)

RESIDENCE: He lived most of his life on Mackinac Island S-1850 US Census
In 1863: Little Traverse (=Harbor Springs), Emmet County S-AGR
DEATH DATE: May 12, 1864 (*Killed in action at Spotsylvania, VA*)
GRAVE: Fredericksburg National Cemetery, Fredericksburg, VA., Grave # 4366
The name on gravestone is "Henry Graverates"
Note: The gravestone for him on Mackinac Island is a memorial stone (See note below)
PENSION FILE: Yes, see Appendix A; 1883 Pension Roll: No
VETERAN CENSUS: 1888 No / 1890 No / 1894 No
RAY HEREK's BOOK: Pages: 36, 59, 148, 150, 155, 344, 389, 434(Roster), 450

SPECIAL INFORMATION:
Henry Graveraet was one of the oldest enlisted men in Company K – and perhaps the wisest. This gruff old fur trader from Mackinac Island was the perfect man to be the First Sergeant of Company K. His many years of working as an interpreter and fur trader taught him to understand the languages spoken by Native Americans. His own native background (¼ Chippewa) and marriage to the granddaughter of a prominent chief taught him respect and understanding for native customs and traditions. He was well known and respected by all the tribes in Northern Michigan. The 2nd Lieutenant of Company K was Garrett Graveraet, his son. Both men died in the war. On Mackinac Island, in St. Anne Cemetery, there are two military gravestones side by side in the family plot, one for Henry, the other for Garrett. However, Henry's stone is only a memorial stone. He is actually buried in Virginia, in Fredericksburg National Cemetery as "Henry Graverates". **His name is on two monuments: Mackinac Island's Veterans Monument, and Traverse City's Veterans Monument in the park at 11th Street and Elmwood Avenue, on a block of stone listing local men who died in the Civil War. Look for the name "H GARRAT GRAVERAT".** According to the "Annual Report to the Commissioner of Indian Affairs 1864", by D.C. Leach, "**Mankewenan**" was the name of Henry Graveraet, the father of Garrett Graveraet. The report also says that Henry was the Chief of the Bear River band. This is an error in a primary source document that is often repeated by modern day historians. Mankewenan is not Henry Graveraet's Anishinabe name, it is a misspelling of Mwakewenah, another soldier in Company K. S-ccn041a
In Marquette, Michigan, Graveraet High School was named after his sister, Juliet. S-ccn041b
The family name, Graveraet, is a Dutch name. Henry's ancestors came from Holland and settled in New York near Albany. They migrated from Albany to Detroit around 1772. S-Shier pg 138

BENJAMIN C. GREENSKY 042

Greensky, Benjamin C., Charlevoix. Enlisted in company K, First Sharpshooters, Sept. 10, 1863, at Little Traverse, for 3 years, age 24. Mustered Oct. 21, 1863. Promoted Corporal. Killed in action at Spottsylvania, Va., May 12, 1864.

Greensky, Benjamin C. – PRIME NAME, Adjutant General Record

RESIDENCE IN 1863: Charlevoix, Charlevoix County S-Descriptive Roll, State Archives

DEATH DATE: May 12, 1864 (*Killed in action at Spotsylvania, VA*)
GRAVE: Location Unknown (*Buried somewhere on the Spotsylvania battlefield*)

PENSION FILE: Yes, see Appendix A; 1883 Pension Roll: No
VETERAN CENSUS: 1888 No / 1890 No / 1894 No
RAY HEREK's BOOK: Pages: 148, 389, 434(Roster)

SPECIAL INFORMATION:
Near Charlevoix, next to Susan Lake, is a place called Greensky Hill. On the hilltop, a Chief who had converted to Christianity started a Protestant mission church. His name was Peter Greensky, and he had two sons named Benjamin and Jacob. The oldest son, Benjamin, loved hunting.
In summertime he would travel to the upper peninsula and live in a hunting camp on Grand Island, near Munsing (S-1860 US Census). He fell in love with a woman living on the Leelanau Peninsula named Emma Redbird. They were married in Northport at the M.E. Church on July 13, 1863 by Rev. Salmon Steele. Eight weeks after the wedding, Benjamin met Garrett Graveraet, who was recruiting volunteers for Company K. After a long discussion, Benjamin said goodbye to his new wife and went off to war. He never returned. Army records tell what happened to him: "Promoted Corporal. Killed in action at Spotsylvania, Virginia, May 12, 1864". Back home, his seventeen year old widow rejoined her family in Northport and applied to the government for a pension. The short marriage produced no children. Benjamin's brother, Jacob, left his home on Greensky Hill and enlisted in Company K. Within weeks he was a soldier in Virginia, fighting alongside the same men his older brother had led into battle. The rest of Jacob's story is told on page 80 of this book.

NOTES FOR HISTORIANS: Ben enlisted in Co. K Sept 10, 1863, eight weeks after his marriage. Greensky Hill was sometimes called Susan Lake Indian Church. It was also called the Pine River Indian Mission, because the nearby town of Charlevoix was originally called "Pine River".
The log-cabin church built on Greensky Hill by Peter Greensky and his two sons, Benjamin and Jacob, still exists! And it is still used as a church. After the Civil War, there were several veterans of Company K in the congregation. Funeral services for these men were held in this log church. Today Greensky Hill and Charlevoix (the city) are in Charlevoix County, but at the time of the Civil War, they were in Emmet County. In 1869, Emmet County was split in two, and the northern half became Charlevoix County. Keep this in mind when searching for the Greensky family in Census Records.

JACOB GREENSKY 043

Greensky, Jacob. Enlisted in company K, First Sharpshooters, Sept. 2, 1864, at Grand Rapids, for 1 year, age 18. Mustered Sept. 2, 1864. Discharged at Alexandria, Va., June 22, 1865.

Greensky, Jacob – PRIME NAME, Adjutant General Record

RESIDENCE IN 1864: Charlevoix, Charlevoix County S-Descriptive Roll, State Archives

DEATH DATE: August 31, 1887 S-pf-043 *(died in Hayes township, Charlevoix County)*
GRAVE: Probably in Greensky Hill Methodist Church Cemetery, Charlevoix Co.
All graves are marked with white crosses, but there are no burial records to show which cross marks his grave

PENSION FILE: Yes, see Appendix A; 1883 Pension Roll: No

VETERAN CENSUS: 1888 No / 1890 Yes(widow) / 1894 No

RAY HEREK's BOOK: Pages: 407, 434(Roster)

SPECIAL INFORMATION:
Near Charlevoix, next to Susan Lake, is a place called Greensky Hill. On the hilltop, a Chief who had converted to Christianity started a Protestant church. His name was Peter Greensky, and he had two sons: Benjamin and Jacob. The three of them built a log-cabin church on the hill. This church is still in existence, and is still used for Sunday services.
During the Civil War, both sons served in Company K of the 1st Michigan Sharpshooters. Benjamin was the first to volunteer. He was promoted to Corporal. Killed in action at Spotsylvania. When news of his death reached Greensky Hill, his brother Jacob went to Grand Rapids and enlisted in Company K. From there he went to Virginia and took his brother's place in the ranks. Jacob Greensky survived the war, came home to Greensky Hill, and the following year married Esther Deverney. One week after their wedding on March 29, 1866, his father, the famous Rev. Peter Greensky, passed away. It appears that Jacob stepped into his father's shoes and conducted services on Greensky hill. But he was not a well man. While in the army he contracted dysentery, and never fully recovered. He lived long enough to see his children grow up and start their own lives. Then, in the summer of 1887, he walked on. Behind the log-cabin church he built with his father and his brother is a quiet cemetery. The graves are marked with white wooden crosses. Jacob Greensky is probably buried under one of those simple crosses, but there are no records to indicate which one.

NOTE TO HISTORIANS:
Today Greensky Hill and Charlevoix (the city) are in Charlevoix County, but during the Civil War they were in Emmet County. In 1869, Emmet County was split in two, the northern half became Charlevoix County. Remember this when searching for the Greensky family in census records.

PETER GRUET 044

Gruet, Peter, Marquette. Enlisted in company K, First Sharpshooters, July 6, 1863, at Dearborn, for 3 years, age 21. Mustered July 11, 1863. Deserted July 19, 1863.

Gruitt, or Greuitt, Peter. Enlisted in company K, Ninth Cavalry, April 24, 1864, at Spaulding, for 3 years, age 24. Mustered June 15, 1864. Joined regiment at Nicholasville, Ky., June 28, 1864. Mustered out at Lexington, N. C., July 21, 1865.

Gruet, Peter – PRIME NAME, from Adjutant General Record

RESIDENCE IN 1863: Marquette, Marquette County S-AGR (1st Mich SS)
In 1864: Spaulding township, Saginaw County S-AGR (9th Mich Cav)
In 1881: Parisville, Huron County (near Ubly) S-pf-050 (pension file for William Isaacs)

DEATH DATE: Unknown *(After 1881, when he gave testimony for William Isaacs)*
GRAVE: Location Unknown *(Last known residence, Parisville in Huron County)*
PENSION FILE: No Pension File exists
VETERAN CENSUS: 1888 No / 1890 No / 1894 No
RAY HEREK's BOOK: Pages: 434(Roster)

SPECIAL INFORMATION:
What happened to Peter Gruet after he deserted in July of 1863 is uncertain.
He appears to have joined another regiment, the 9th Michigan Cavalry, and served till the end of the war, receiving an honorable discharge. But that is not a proven fact.
To anyone who wants to solve the mystery of Peter Gruet, here are five clues:
Clue #1 In the history of Clinton County, a man named James Gruet was an interpreter at the Bethany Indian mission. Some records indicate he had a grandson named Peter Gruet Jr. who was born about the same time as the soldier in Company K. S-ccn044a
Clue #2 Official army records (Descriptive Roll, State Archives) say "Peter Gruet" enlisted in the 1st Michigan Sharpshooters, Company K, at Dearborn. He told the army his residence was Marquette in the Upper Peninsula. The same army records say Peter was mustered in on July 11, and deserted 8 days later on July 19.
Clue #3 The records of the 9th Michigan Cavalry say a man named Peter Gruitt / Greuitt enlisted into that regiment in April, 1864, at Spaulding, Michigan (Spaulding township, Saginaw County). He survived the war and received an honorable discharge. This 9th Cav soldier has no pension file.
Clue #4 The pension file of Company K soldier Thomas Nelson (of Taymouth, Saginaw Co.) says the legal guardian of Nelson's children was a man named Peter Grewett Jr. S-pf-086 (Nelson pension)
Clue #5 In 1881, a man named Peter Grouett living near Parisville (Huron Co., near Ubly) appeared before a Saginaw County Clerk and gave sworn testimony for Company K soldier William Isaacs, saying he knew William well during the ten year time period before he (William) joined the army, and saw him when he came back from the war. S-pf-050 (pension file for William Isaacs)

CORNELIUS HALL 045

Hall, Cornelius. Enlisted in company K, First Sharpshooters, Sept. 3, 1864, at Grand Rapids, for 1 year, age 18. Mustered Sept. 3, 1864. Discharged at Alexandria, Va.

Hall, Cornelius – PRIME NAME, Adjutant General Record
Hall, Corneilius W. – State Death Certificate name
Ne-gaw-ne-say – His Odawa name S-Durant Roll Field Notes #18-pg60

RESIDENCE IN 1860: Pere Marquette township, Mason County S-1860 US Census
In 1894: P.O. Star City, West Branch township, Missaukee County S-1894 Veteran Census

DEATH DATE: August 5, 1917 (*Died in West Branch twp., Missaukee County, Michigan*)
S-State Death Record

GRAVE: Missaukee County / Star City Cemetery (Military stone)
(*Location: Star City Cemetery is 20 miles straight west of Houghton Lake*)
To find the cemetery: Start at intersection of US-127 and M-55, go west on M-55 to Star City road, turn North and travel on Star City road a few miles. The cemetery will appear on the left side (the west side) of the road. Look for Hall's grave in the back of the cemetery, in the Indian section.

PENSION FILE: Yes, see Appendix A; 1883 Pension Roll: No

VETERAN CENSUS: 1888 No / 1890 Yes / 1894 Yes

RAY HEREK's BOOK: Pages: 407, 434(Roster)

SPECIAL INFORMATION:
He was born the son of an Ottawa Chief. S-ccn045a
His father was Kaw-gay-gaw-bo-we, Chief of one of the Grand River bands of Ottawa who are known today as the Little River Band of Ottawa Indians. S-ccn045a
His Anishinabe name was **Ne-gaw-ne-say**. S-Durant Roll Field Notes #18-pg60
Cornelius grew up on a farm near Ludington in Mason County. S-1860 US Census
He enlisted in Company K near the end of the war as a replacement and arrived just in time to take part in the last battles of the Siege of Petersburg. Wounded in action April 2. S-pf-045
Discharged May 20, 1865, from an army hospital in Alexandria, VA. S-pf-045
After the war he returned home, married, and settled down to the life of a farmer in Mason Co. Around 1885 he moved to Missaukee County, bought a farm near Star City, and lived there till his death in 1917. S-1880 US Census, and the 1890 Veteran Census
Cornelius was member # 54 of the Caldwell G.A.R. Post No. 365, Lake City, Missaukee Co.
His name appears on the Veterans Monument in Lake City (corner of Main and Prospect).
There is a photograph of Cornelius Hall and his family taken in 1903. You can find it in the collection of the Wexford County Historical Society and Museum.

JAMES H. HAMLIN 046

Hamblin, James M. Enlisted in company K, First Sharpshooters, March 28, 1864, at Mackinac, for 3 years, age 22. Mustered April 29, 1864. Joined regiment at North Anna river, Va., May 26, 1864. Taken prisoner at Petersburg, Va., June 17, 1864. Died in prison Oct. 21, 1864. Buried in National Cemetery at Andersonville, Ga. Grave No. 11260.

Hamlin, James H. – PRIME NAME, from Pension File
Hamblin, James M. – Adjutant General Record name
Hamelin, Jacques Hubert – His name when baptized S-Records of St. Anne Church

RESIDENCE IN 1864: St. Ignace, Mackinac County S-Descriptive Roll, State Archives
DEATH DATE: October 20, 1864 (*Died while a POW at Andersonville Prison*) S-AGR
(*Adjutant General says October 21. The Atwater list says October 20. Always trust Atwater.*)
GRAVE: Georgia / Andersonville National Cemetery, Grave # 11,260
GPS N32.20470 W84.13074
PENSION FILE: Yes, see Appendix A; 1883 Pension Roll: No
VETERAN CENSUS: 1888 No / 1890 No / 1894 No
RAY HEREK's BOOK: Pages: 398, 413, 434(Roster)

SPECIAL INFORMATION:
The correct spelling of his name is James H. Hamlin.
There is no B in his name, and his correct middle initial is "H", not "M". S-pf-046
He was born and raised on Mackinac Island. S-1850 US Census, head of house "James Hamlin"
By 1860 he had moved off the Island to St. Ignace, a married man with a family. S-1860 US Census
Married Rosalie Jeandron on October 2, 1855. Rosalie is on Durant Roll, see Field Notes #7-pg15.
His correct age at enlistment was 42 (b. Feb 12, 1822). He enlisted in the army on Mackinac Island, but resided on Point St. Ignace with his wife and 5 children. Occupation: Fisherman. S-pf-046
Member of the Mackinac Band, and the brother-in-law of Co. K soldier Michael Jeandron.S-ccn046a
James and his brother in law Michael enlisted in the army together, fought in battle together, were captured together, and died together at the infamous Andersonville prison camp.

NOTES FOR HISTORIANS:
To anyone researching the French in Michigan, the family of James Hamlin is a fascinating study. He was baptized "Jacques Hubert Hamelin" in St. Anne Church on Mackinac Island by the famous Father Gabriel Richard. His brother was Augustin Hamelin Jr. (aka Kiminitchagan) who traveled to Rome (a great story). His father was Augustin Hamelin Sr. According to family tradition, his grandfather, Louis Hamelin, was a French soldier who fought under Charles Langlade in the French & Indian war and was present at the surrender of Quebec in 1759. And his great-grandfather was a French soldier named Charles, who transferred to Fort Michilimackinac from a post in Quebec. His full name was Charles-Francois Hamelin-LaGuenier, born 1714 in Grodines, Quebec. We know exactly where he lived inside Fort Michilimackinac, because his residence is marked on the 1749 "Lotbiniere Map" of soldier residences, with the name "Charles Amlin".

JOSEPH HANNIN 047

Hannins, Joseph. Enlisted in company K, First Sharpshooters, Feb. 3, 1865, at Mendon, for 3 years, age 25. Mustered Feb. 4, 1865. Mustered out at Delaney House, D. C., July 28, 1865.

Hannin, Joseph – PRIME NAME, from Pension File
Hannins, Joseph – Adjutant General Record name
Hannius, Joseph – Adjutant General Annual Report
Honyust, Joseph – name on his Death Record.
Harniost, Joseph – name on the 1851 Canadian Census

RESIDENCE IN 1865: Mendon, St. Joseph County S-AGR
1830: Oneida Castle, New York (an Oneida village) S-pf-047
1851: The Oneida Settlement in Middlesex County, Ontario, Canada S-Census Record
DEATH DATE: January 7, 1916 S-pf-047
Died on the Oneida Settlement (Reserve 41) Middlesex Co., Ontario, Canada) S-pf-047
GRAVE: Center Church Cemetery, Oneida Settlement (Reserve 41), in Middlesex County, Ontario, Canada S-County Death Record (Middlesex County, Ontario, Canada)
Center Church Cemetery is on Fairgrounds Road, north of the intersection with Oneida Road. There is no gravestone, and burial records for this cemetery have not yet been located.
PENSION FILE: Yes, see Appendix A; 1883 Pension Roll: No
VETERAN CENSUS: 1888 No / 1890 No / 1894 No
RAY HEREK's BOOK: Pages: 434(Roster), 496

SPECIAL INFORMATION:
Joseph Hannin (aka Harniost and Honyust) was an Iroquois born in the state of New York at Oneida Castle on February 1, 1830 (S-pf-047). When he was a boy, his tribe sold their land to settlers and moved to Canada. They bought new land in Middlesex County, Ontario. Today his people are known as the Oneida Nation of the Thames, and their reservation is named Oneida Reserve 41, aka "The Oneida Settlement". His name appears on Canadian census records as "Joseph Harniost". For unknown reasons, in 1865 Joseph left his reservation in Canada to live in Mendon, Michigan. He enlisted in Company K just in time for the final assault on Petersburg. After his discharge from the army, Joseph Hannin returned to Canada and lived on Oneida Reserve No. 41 in Ontario for the rest of his life. He is buried there in Center Church Cemetery. S-Death Record
Joseph Hannin was one of two Iroquois men in Company K. The other was David Lidger. Both men were born in Oneida, New York, and enlisted together in Company K.

NOTES FOR HISTORIANS:
Joseph claimed to be 25 when he enlisted, but records indicate he was actually 35 years old. In Canada, an Indian Reservation is called an Indian Reserve, and each one has an ID number. The "Oneida Settlement" (Reserve 41) is in Delaware township, Middlesex County, Ontario.

CHARLES HUBERT 048

Hubert, or Herbert, Charles, Sault Ste. Marie. Enlisted in company K, First Sharpshooters, Aug. 9, 1863, at Detroit, for 3 years, age 28. Mustered Sept. 29, 1863. In Mower Hospital, Philadelphia, Pa., Jan. 6, 1865. No further record.

Hubert, Charles – PRIME NAME, Adjutant General Record
Herbert, Charles – Adjutant General Record, aka name
Hubbard, C. – Name on records of Mower Hospital, Philadelphia

RESIDENCE IN 1863: Sault Ste. Marie, Chippewa Co. (Upper Peninsula) S-AGR

DEATH DATE: Unknown

GRAVE: Location Unknown

PENSION FILE: No Pension File exists

VETERAN CENSUS: 1888 No / 1890 No / 1894 No

RAY HEREK's BOOK: Pages: 401, 434(Roster)

SPECIAL INFORMATION:
He was wounded in action on July 20, 1864 at Petersburg, VA. S-Herek pg 401
It appears his wound was serious, because six months later he was still in a hospital. S-AGR
Last record is dated Jan 6, 1865, when he was a patient at Mower Hospital, Philadelphia, under the name "C. Hubbard". Then he disappears. S-Descriptive Roll, State Archives

NOTES FOR HISTORIANS:
Anyone trying to solve the mystery of what happened to Charles Hubert should study the case of Thomas Miller, another soldier in Company K staying at the same hospital who disappears from the records. We know Miller left Mower hospital on convalescent leave in September 1864 and died at home (S-pf-076, Miller pension file). Yet his "official" army record says he deserted on June 6, 1865, eight months after he died! I suspect that Charles Hubert was also sent home to Michigan on convalescent leave, and like Miller he did not return. Further research needed.
There was an Ojibwa Metis named Charles Hubert born in St. Ignace in 1834, son of Jack Hubert, nephew of the Chief of Grand Traverse. The 1836 Mixed Blood Census Register, connected with the Treaty of March 28, 1836, says the following: Charles Hubert, Ojibwa Metis, born 1834 at St. Ignace, was the son of Jack Hubert, an Ojibwa Metis, and an Ojibwa Metis Woman.
Is this man the same Charles Hubert who enlisted in Company K? More research needed.

JOHN ISAACS 049

Isaacs, John, Isabella. Enlisted in company K, First Sharpshooters, May 19, 1863, at Port Huron, for 3 years, age 21. Mustered May 26, 1863. Joined regiment at Peebles' Farm, Va., Nov. 4, 1864. Discharged at Detroit, Mich., Aug. 11, 1865.

Isaacs, John – PRIME NAME, from Adjutant General Record
Isaacks, John – Adjutant General Annual Report

RESIDENCE IN 1863: Isabella, Isabella County (Enlisted in Port Huron) S-AGR

DEATH DATE: Unknown

GRAVE: Location Unknown

PENSION FILE: No Pension File exists

VETERAN CENSUS: 1888 No / 1890 No / 1894 No

RAY HEREK's BOOK: Pages: 272, 273, 434(Roster)

SPECIAL INFORMATION:
There were two soldiers in Company K with last name Isaacs: John and William. According to Ray Herek's book, these men were brothers. S-Herek pg 272
The two brothers were from Isabella, but they enlisted on different dates at different places. Why John Isaacs enlisted in Port Huron is a mystery. John deserted for a while but rejoined the regiment on Nov. 4, 1864. There was no court martial, and he received an honorable discharge at the end of the war. S-Herek pg 272
Most of Company K was discharged July 28, 1865 in Washington D.C. Notice that John Isaacs was discharged August 11, 1865 in Detroit. This later date suggests that at the end of the war he was in a hospital, seriously ill. When he became well enough to travel, the army sent him home. What happened to him after the war is a mystery.

NOTE TO HISTORIANS: John Isaacs may be the mysterious John Michigan, a Native American veteran of the Civil War buried in a solitary grave hidden deep in the forests of Iosco County. Local legend says two Indians from Au Sable went to fight in the Civil War. Both survived, but on the way home one was robbed and killed. It is assumed this is the grave of the one who came home. The grave is near the Oscoda Methodist Indian Mission, on the corner of King's County Road (aka County Line Highway) and Anvil Road, in Iosco County. See the book "IOSCO'S BOYS IN BLUE, A ROSTER OF RESIDENT CIVIL WAR SOLDIERS, SAILORS, AND VETERANS", published 1993 by James Donaldson. On Pg 38 is the story of John Michigan.

WILLIAM ISAACS 050

Isaacs, William, Isabella. Enlisted in company K, First Sharpshooters, May 11, 1863, at Isabella, for 3 years. Mustered May 26, 1863. Wounded in action June 17, 1864. Mustered out at Delaney House, D. C., July 28, 1865.

Isaacs, William – PRIME NAME, from Adjutant General Record
Isaac, William – Pension File name
Isiacks, William – State Death Record name

RESIDENCE IN 1863: Isabella township, Isabella County S-AGR
In 1881: Taymouth, Saginaw County S-pf-050
In 1890 & 1894: Nunda township, Cheboygan County S-1890 and 1894 Veteran Census
In 1907: Living in Bay Shore, Charlevoix County, at the time of his death S-ccn050a
DEATH DATE: March 16, 1907 (*Died in Boyne City, Charlevoix Co.*) S-State Death Record
For details of his death, see story in the Muskegon Daily Chronicle, Tuesday March 26, 1907, page 1, Column 3
GRAVE: Charlevoix County, Boyne City / Maple Lawn Cemetery (Military stone)
Buried in Potter's Field, plot No. 89.
PENSION FILE: Yes, see Appendix A; 1883 Pension Roll: No
VETERAN CENSUS: 1888 No / 1890 Yes / 1894 Yes
RAY HEREK's BOOK: Pages: 272, 396, 434(Roster)

SPECIAL INFORMATION:
John Isaacs and William Isaacs were brothers, according to Ray Herek's book. S-Herek pg 272
William's eyes were injured, and how it happened is a story worth telling – He was sitting on the ground during a battle when a cannon ball fell from the sky and landed right between his legs. Missed him by an inch! However, flying dirt was driven into his eyes. His eyesight deteriorated steadily after that incident. S-pf-050
The children of William Isaacs saved his original discharge paper and passed it down in the family. The author has been allowed to examine it and have it professionally copied. It is the only known Discharge Paper for a Company K soldier in existence. S-ccn050b
This discharge paper says he was 20 yrs. old when he left the army in 1865 [born 1845].
William was a charter member of G.A.R. Post No. 264 in Wolverine, Cheboygan County.
He was buried in plot 89 of Potter's Field of Maple Lawn cemetery in Boyne City. The grave was unmarked until 2011, when the Sons of Union Veterans ordered a Military Gravestone for it.
NOTES TO HISTORIANS:
There is no doubt that William Isaacs was wounded in battle, but where it happened and how it happened is unclear. Official Army Records say he was wounded June 17, 1864. However, his pension file says he was shot through the right leg near the knee at White House Landing around June 11, 1864. A report from the Surgeon General indicates he was in the hospital on June 17, so he could not possibly have been in a battle on that day. S-pf-050
The correct spelling of his name is a problem. William consistently used "Isaac" with no s. His granddaughter used "Issacs". Army records & his brother spelled it "Isaacs", with an s. For his prime name, I choose the spelling used by the Army records and his brother.

JOHN JACKO 051

Jacko, John. Enlisted in company K, First Sharpshooters, Feb. 14, 1865, at Grand Rapids, for 1 year, age 20. Mustered Feb. 14, 1865. Mustered out at Delaney House, D. C., July 28, 1865. Present residence, Horton's Bay, Mich.

Jacko, John – PRIME NAME, Adjutant General Record
Jaco / Iaco, John – Name on enlistment papers S-pf-051
Shawkoo, John – Name on Durant Roll S-Durant Roll No. 6315, Field Notes #11-pg39
Kaw-pon-kay – Name on 1870 Indian Annuity S-Durant Roll Field Notes #11-pg39

RESIDENCE IN 1865: Northport, Leelanau County S-pf-051
In 1890: Horton Bay, Evangeline township, Charlevoix County S-1890 Veteran Census
In 1894: Hayes township, Charlevoix County S-1894 Veteran Census
DEATH DATE: April 21, 1907 (*Died in Evangeline township, Charlevoix Co.* S-ccn051a)
(*Note: Pension file has a County Death Certificate saying he died April 4. This is wrong* S-ccn051b)
GRAVE: Charlevoix County, Boyne City / Maple Lawn Cemetery (Military stone)
He is buried in plot No. 268. (*Look for his grave near a tall statue of a Civil War soldier)*
PENSION FILE: Yes, see Appendix A; 1883 Pension Roll: No
VETERAN CENSUS: 1888 No / 1890 Yes / 1894 Yes
RAY HEREK's BOOK: Pages: 290, 434(Roster)

SPECIAL INFORMATION:
John Jacko was born and raised in Northport, Leelanau County S-pf-051
The family name is also spelled Shako. There are many Shakos in the Peshawbestown area.
The following three members of his family enlisted in Company K:
First, his father Jacko Penaiswanquot. After the father died in Andersonville Prison, his two oldest sons enlisted in Company K. There is no evidence saying they enlisted to avenge their father. The sons were John and Louis. (Louis was also called Nat-tah-me-no-ting). S-pf-051
Horton Bay was John's home. This town is famous for its association with Ernest Hemmingway.
John was a member of the Miles Norton G.A.R. Post No. 401, Horton Bay, Charlevoix Co.
Three photographs of John Jocko exist in the collection of the Charlevoix Historical Museum.
When John Jacko died, his obituary was on Front Page of the April 25, 1907 Charlevoix Sentinel.
John Jacko's grave was unmarked till the Sons of Union Veterans ordered a gravestone in 2011.
The new stone was dedicated on May 7, 2011. The Jacko family attended the ceremony.
There is a gravestone for John Jacko in Elmwood Cemetery, Detroit, and it is a mistake.
It rests on the grave of his brother Louis, aka Nat-tah-me-no-ting Jacko. S-ccn051c

NOTE TO HISTORIANS:
His pension file contains a death certificate that says he died April 4. This is wrong.
Correct death date is April 21, 1907. S-State Death Certificate and newspaper obituary
Durant Roll Field Notes #11-pg39 say he died April 22; correct death date is April 21.

NATTAHMENOTING JACKO 052

(This soldier does not appear in the Adjutant General Record)

Jacko, Nat-tah-me-no-ting – PRIME NAME, taken from a book titled Father Abraham's Children by Woodford. Pgs. 212-213
Jacko, Nah-tah-mi-no-ting – Cemetery burial records
Nat-bah-me-no-ling, Jacko – Gravestone name (old gravestone)
Penaiswanquot, Louis – Pension File name
Natawinoding, Louis – Pension File, aka name

RESIDENCE IN 1865: Northport, Leelanau County S-pf-098 (Penaiswonquot pension file)
DEATH DATE: November 28, 1870 S-burial records in the office of Elmwood cemetery
GRAVE: Wayne County, Detroit / Elmwood Cemetery, Section S (Military stone)
Section S, Lot 14, Row 3. Note: Section S is a Soldiers Section. There is a flagpole in front of it, making it easy to find. Jacko's grave is in the left half, in the third row.
PENSION FILE: Yes, see Appendix A (two files) ; 1883 Pension Roll: No
VETERAN CENSUS: 1888 No / 1890 No / 1894 No
RAY HEREK's BOOK: No mention of this man

SPECIAL INFORMATION:
There is more confusion and mystery about this soldier than any other man in Company K. Here are the facts: He is buried in Elmwood Cemetery, Detroit, with a military gravestone. His old gravestone says Company K, 1st Michigan Infantry (not 1st Sharpshooters). His new gravestone says "John Jacko, Company K, 1st Michigan Sharpshooters, 1841-1870". This is completely wrong because John Jacko was alive and well till 1907, and is buried in Boyne City. **Two pension files exist for this soldier: He applied for an Invalid pension in 1870 when he was in the hospital (Application No. 158,777). Years later, in 1896, his mother applied for a Mother's pension (Application No. 645,502. Note: her application papers and testimony proving Louis was her son are filed in her widow pension file, Cert. No. 263,983).**
This man was born Louis Penaiswanquot, son of Jacko Penaiswonquot & his wife Mary. His Anishinabe name was Natawinoding (=Fresh Wind). His father joined Company K and died in Andersonville prison. After the father was captured, his oldest sons, Louis and John, enlisted in Company K. They did not enlist together. John Jacko is on the roster. Louis is NOT, and the reason why is a mystery. **Perhaps Louis is on the roster under the name Jacob Prestawin?** Same enlistmen date, place, and physical description. Further research needed. **His pension file tells this story: Louis enlisted November 18, 1864 at Traverse City for one year. Joined his unit at Petersburg and fought in the last battles. Nearly died from typhoid fever. After the war ended, he was discharged from a hospital in Washington DC on July 3, 1865.**
We know for certain he died in 1870 of Consumption (tuberculosis) which he claimed was contracted while in the army. The book Father Abraham's Children, by Frank B. Woodford, says he died on November 29, 1870 in Harper Hospital, Detroit. See pages 212-213 for details. Death Date Controversy: Woodford's book says he died November 29, but cemetery records say Nov. 28. **His last name should be Jacko, because the other sons of Jacko Penaiswanquot used Jacko as their last name. Also, the Woodford book says his last name is Jacko.**

EDWARD A. JACKSON 053

Jackson, Andrew. Enlisted in company K, First Sharpshooters, Oct. 1, 1864, at Detroit, for 3 years, age 27. Mustered Oct. 1, 1864. Joined regiment at Peebles' Farm, Va., Nov. 4, 1864. Discharged at Washington, D. C., June 15, 1865.

Jackson, Edward Andrew – PRIME NAME S-Death Certificate in his pension file
Jackson, Andrew – Adjutant General Record & Pension File
Scholes, Edward – 1888 Veteran Census

RESIDENCE IN 1864: Salisbury, Massachusetts S-ccn053a
Born July 4, 1837 (or 1834) in Hyde Park, New York (Dutchess County) S-ccn053a
In 1864: Living with father in Salisbury, Massachusetts S-ccn053a
In the 1880's: Living in Blenheim, Kent County, Ontario (Canada) S-ccn053a

DEATH DATE: December 23, 1893 (*Died in Detroit, Wayne Co., Michigan*) S-ccn053b
GRAVE: Canada, Ontario, Kent Co., Blenheim / Evergreen Cemetery (Civilian stone)
PENSION FILE: Yes, see Appendix A; 1883 Pension Roll: No
VETERAN CENSUS: 1888 Yes / 1890 No / 1894 No
RAY HEREK's BOOK: Pages: 434 (Roster)

SPECIAL INFORMATION:
According to army records, his name was Andrew Jackson. But Andrew was his middle name. His full, legal name was **Edward Andrew Jackson**. Genealogy records and his descendants give this story: He was born in Hyde Park, Dutchess Co., New York. His father was white, his mother was full-blood Indian. He and his father lived at one time in Salisbury, Massachusetts. He was living in Harwich, Ontario (Kent Co., Canada) when he traveled to Detroit to enlist in Company K. Almost died of Typhus fever, but survived. Discharged from an army hospital.
After the war he returned to Harwich, Ontario. There he married and raised a family.
The 1888 Veteran Census has a member of Company K named "Edward Scholes" living in Detroit. I believe this is Edward A. Jackson. Some of his children moved to Detroit; maybe they were taking care of him. Where the name Scholes comes from is a mystery – Bad handwriting ?
Edward Andrew Jackson belonged to the IOF (Independent Order of Foresters, a Canadian Fraternal Organization) and the R T of T (Royal Templars of Temperance).
His unusual gravestone, a cement tree stump, was probably provided by the IOF.
Edward was born on the 4th of July, but the exact year is uncertain. Most sources say 1837, and that is probably correct. His gravestone says he was born 1839, but that could be an error.
He had a son who enlisted in the army in 1898 during the Spanish American war. You can see the son's name on a bronze tablet in Ypsilanti honoring local men who joined Co. G of the Michigan 31st Infantry. His name is Edward Andrew Jackson, same as his father.

WILLIAM JACKSON 054

Jackson, William. Enlisted in company K, First Sharpshooters, Aug. 29, 1864, at Flint, for 3 years, age 26. Mustered Aug. 29, 1864. Substitute for Newell Barnard. Mustered out at Delaney House, D. C., July 28, 1865.

Jackson, William – PRIME NAME, all records agree on spelling

RESIDENCE IN 1864: Saginaw, Saginaw County S-Descriptive Roll, State Archives

DEATH DATE: Unknown

GRAVE: Location Unknown

PENSION FILE: No Pension File exists

VETERAN CENSUS: 1888 No / 1890 No / 1894 No

RAY HEREK's BOOK: Pages: 250, 434(Roster)

SPECIAL INFORMATION:
William Jackson and George Kabayacega enlisted in Flint on the same day. Both men were from Saginaw County. Draft credit was given to the city of Saginaw. When the army asked him his occupation, he replied "Hunter". S-Descriptive Roll, State Archives
He enlisted in the later part of the war as a substitute for Newell Barnard, a man who was apparently going to be drafted. Barnard paid Jackson to take his place. William Jackson was lucky in the war - never wounded or seriously ill. After Lee surrendered, he was discharged with the rest of Company K …and disappears from all records.
What happened to him after he left the army is a mystery.

MICHAEL JEANDRON 055

Jondrau, Michael. Enlisted in company K, First Sharpshooters, March 14, 1864. Joined regiment at North Anna River, Va., May 26, 1864. Taken prisoner in action at Petersburg, Va., June 17, 1864. Died in prison Jan. 5, 1865. Buried at Andersonville, Ga.

Judson, Michel. Enlisted in First Sharpshooters, Feb. 29, 1864, at Mackinaw, for 3 years, age 21. Mustered April 29, 1864. No further record.

Jeandron, Michael – PRIME NAME, Adj. General Annual Report
Jondrau, Michael – Adjutant General Record name
Jourdron, M – Gravestone name at Andersonville National Cemetery
Jondro, M – Atwater List name S-ccn055a
Judson, Michel – Misspelling of the name Jeandron, in my opinion

RESIDENCE IN 1864: St. Ignace, Mackinac Co. S-Descriptive Roll for Michael Jondrau
DEATH DATE: January 5, 1865 (*Died while a POW at Andersonville Prison*) S-AGR
GRAVE: Georgia / Andersonville National Cemetery, Grave No. 12,396
GPS N32.20481 W84.13049
PENSION FILE: No Pension File exists
VETERAN CENSUS: 1888 No / 1890 No / 1894 No
RAY HEREK's BOOK: Pages: 398, 413, 434(Roster)

SPECIAL INFORMATION:
Michael Jeandron's service record is short and sad. Three weeks after he arrived at the camp of Company K in Virginia, he was captured, sent Andersonville prison, and died of starvation and disease.
He enlisted in the army on Mackinac Island, but his home was in St. Ignace. S-Descriptive Roll
He was a member of the Mackinac Band, and was the brother-in-law of James Hamlin. S-ccn055b
His sister, Rosalie, married James Hamlin. Michael and James enlisted into the army together, fought in battles together, were captured together, and died at Andersonville prison together.

NOTE TO HISTORIANS:
After examining all available information on Michael Jondrau and Michel Judson, I have come to the conclusion they are the same man. Here are my reasons:

1. Michael Jondrau has no enlistment information (except a date) but has detailed information after he enlisted. Michel Judson has detailed enlistment information, but no information after he enlisted. Combine these two records, and you have one complete service record.
2. Similarity of names. Same first name, last names are so close that bad handwriting could explain the difference in spelling.
3. The Adjutant General Annual report gives his name as "Michael Jeandron". Historically, a Michael Jeandron Sr. lived in St. Ignace (he was the grandfather of this soldier). S-ccn055c

 All Rights Reserved by Chris Czopek

DAVID JOHNS 056

Johns, David. Enlisted in company K, First Sharpshooters, Jan. 27, 1865, at LaGrange, for 3 years, age 19. Mustered Jan. 28, 1865. Mustered out at Delaney House, D. C., July 28, 1865.

Johns, David – PRIME NAME, Adjutant General Record
John, David – Pension File name
Waw-bo-ge-maw, David – His native name S-Gruett Roll, pg. 349
Kawis, Mary – His widow's name

RESIDENCES:
Born in Bay City c. 1846 S-pf-056
1865: La Grange township, Cass County S-AGR (see note below)
1870: Kawkawlin township, Bay County S-1870 US Census, family of Jack Fisher
From 1871-1889: Fountain and Custer, Mason County S-pf-056
DEATH DATE: February 28, 1889 (*Died in an accident, see note below)*
GRAVE: Location Unknown (*Died near Fountain in Mason County*)
PENSION FILE: Yes, see Appendix A; 1883 Pension Roll: No
VETERAN CENSUS: 1888 No / 1890 Yes (widow) / 1894 No
RAY HEREK's BOOK: Pages: 434(Roster), 496

SPECIAL INFORMATION:
This soldier was named David Johns. Be careful not mix him up with another soldier in Company K named John David.
Army records say his last name was "Johns" with an s. But his pension file says his last name was John, with no s. Further research needed to determine which spelling is correct.
His native name was David Waw-bo-ge-maw. S-Gruett Roll, pg. 349, son of Jack Waw-bo-ge-maw.
He joined the army in the last months of the war. Fought in the last big battles around Petersburg, VA. When the war ended, he marched in the big victory parade in Washington D.C.
Death Date Controversy: His pension file says he was run over by a train on March 2, 1889, and died a few days later on March 4. No story of this train accident appears in any local newspaper. However, railroad records say he died Feb 28, 1889. Which date is correct? I choose February 28.
His widow, Mary Kawis, applied for a pension after his death.
Two children: Joseph John (b. Sept 1, 1881) and Nancy John (b. March 5, 1889).
David Johns is probably related to a soldier in Company K named William Johns. They both enlisted on the same day (January 27, 1865) and have the same last name (Johns).

NOTES FOR HISTORIANS: He claimed in 1865 to be a resident of LaGrange, Cass County, but it is doubtful that he actually lived there. He was born in Bay City (S-pf-056) and after the war lived with his father in Kawkawlin (S-1870 US Census, family of Jack Fisher). He is definitely a Chippewa. The GRUETT ROLL, page 349, shows the family of a Chippewa named Jack Waw-bo-ge-maw. Jack had two sons, David and William. There is good reason to believe these sons are Company K soldiers David Johns and William Johns. However, further research is needed to prove this fact.

All Rights Reserved by Chris Czopek

WILLIAM JOHNS 057

Johns, William. Enlisted in company K, First Michigan Sharpshooters, Jan. 27, 1865, at White Pigeon, for 3 years, age 20. Mustered Jan. 28, 1865. Mustered out at Delaney House, D. C., July 28, 1865.

Johns, William – PRIME NAME, all sources agree with spelling
Waw-bo-ge-maw, William – S-Gruett Roll, pg. 349, and see note below

RESIDENCE IN 1865: White Pigeon, St. Joseph County S-AGR (See note below)
1870: Kawkawlin township, Bay County S-1870 US Census, family of Jack Fisher

DEATH DATE: 1886-1889 (See note below)

GRAVE: Location Unknown

PENSION FILE: No Pension File exists

VETERAN CENSUS: 1888 No / 1890 No / 1894 No

RAY HEREK's BOOK: Pages: 434(Roster), 496

SPECIAL INFORMATION:
When William Johns joined the army, he said his residence was White Pigeon, St. Joseph County. But there is some doubt about whether or not he actually lived there. He may have given this residence so his enlistment would be credited to the St. Joseph County draft quota.
He enlisted during the last months of the war. Ten weeks after he enlisted, Lee surrendered and William Johns went home.

After the war he lived with his father in Kawkawlin twp., Bay Co. S-1870 US Census, Jack Fisher family
In 1916, a report was made of the heirs of William's mother, Julia Paundegaykawwaw, aka Nokechewawno, aka Julia Francis. It mentions her son William Waw-bo-ge-maw, and says he died c. 1886-1889 and was not married.

NOTE TO HISTORIANS:
William Johns is probably related to a soldier in Company K named David Johns.
They both enlisted on the same day (January 27, 1865) and have the same last name (Johns).
The GRUETT ROLL, page 349, shows the family of a Chippewa named Jack Waw-bo-ge-maw. He has two sons named David and William. There is good reason to believe these sons are Company K soldiers David Johns and William Johns. If true, then William John's native name was **William Waw-bo-ge-maw,** and he was a Chippewa. However, it is not yet a proven fact that William Johns is the brother of David Johns. Further research is needed.

LOUIS KABAOSA 058

Ka-ba-o-sa, Louis, Pentwater. Enlisted in company K, First Sharpshooters, July 4, 1863, at Elbridge, for 3 years, age 27. Mustered July 11, 1863. Mustered out at Delaney House, D. C., July 28, 1865.

Ka-ba-o-sa, Louis – PRIME NAME, from Adjutant General Record
Kubaosa, Lewis – Adjutant General Annual Report
Kabaosa, Louis – Oceana County Native American Death Index 1867-1875

RESIDENCE IN 1863: Pentwater, Oceana County S-AGR

DEATH DATE: July 9, 1870 (*Died of Consumption*) S-ccn058a

GRAVE: Location Unknown (*Died in Elbridge Township, Oceana County*)

PENSION FILE: No Pension File exists

VETERAN CENSUS: 1888 No / 1890 No / 1894 No

RAY HEREK's BOOK: Pages: 434(Roster)

SPECIAL INFORMATION:
When Company K was being formed, Captain Andress made a trip to a small reservation near Pentwater on the 4th of July. He gave a speech to the tribe, asking for volunteers to fight in the Civil War. Twenty five men stepped forward. Louis Kabaosa was one of those volunteers. The full story of the 4th of July at the reservation is on page 35 of Ray Herek's book. When the war ended, Louis Kabaosa went home to Michigan, settled in Elbridge township, and became a farmer. He died of Consumption (Tuberculosis) at age 34 in 1870.

 All Rights Reserved by Chris Czopek

GEORGE KABAYACEGA 059

Ka-ba-ya-ce-ga, George. Enlisted in company K, First Sharpshooters, Aug. 29, 1864, at Flint, for 3 years, age 30. Mustered Aug. 29, 1864. Substitute for John L. Barnard. Mustered out at Delaney House, D. C., July 28, 1865.

Ka-ba-ya-ce-ga, George – PRIME NAME, Adjutant General Record
Kabayahsega, Geo. – Name on gravestone
Keabuorga, George – History of Saginaw County, pg 368 S-ccn059a
Sock-a-tup, George – Pension File, aka name
Sacketup, Gorge – 1890 Veteran Census (widow Mary Sacketup)

RESIDENCE IN 1864: Saginaw, Saginaw County S-Descriptive Roll, State Archives
In the 1870's: St. Charles area, Saginaw County S-pf-059

DEATH DATE: March 30, 1882 S-pf-059
(*Died near St. Charles, Saginaw County, in an accident - See his pension file for details*)

GRAVE: Saginaw County, St. Charles / Riverside Cemetery (Military stone)
Note: Highway M-52 divides the cemetery in half; the grave is located in the east half.

PENSION FILE: Yes, see Appendix A; 1883 Pension Roll: No

VETERAN CENSUS: 1888 No / 1890 Yes (Mary Sacketup, widow) / 1894 No

RAY HEREK's BOOK: Pages: 250, 434(Roster)

SPECIAL INFORMATION:
George Kabayacega and William Jackson enlisted in Flint on the same day. Both men were from Saginaw County. Draft credit was given to the city of Saginaw. Both men enlisted as substitutes. George was paid to be a substitute for John L. Barnard, a man who was going to be drafted. During the war, an exploding shell threw dirt into his eyes. From then on, his eyesight deteriorated. James Fisher, Chief of the Taymouth band, testified that before the war, George had excellent eyesight and earned his living as a hunter (S-pf-059). After the war, his eyesight was so poor, the only way he could earn money was making baskets and axe handles. S-pf-059
He changed his name after the war to George Sacketup / Sagatup / Sock-a-tup. S-pf-059
His widow was known as Mary Sacketup.
After the war, George lived and died in the St. Charles area of Saginaw County.

JOSEPH KADAH 060

Kadah, Joseph, Pentwater. Enlisted in company K, First Sharpshooters, Sept. 9, 1863, at Elbridge, for 3 years, age 21. Mustered Sept. 29, 1863. Died at St. Mary's Hospital, Detroit, Mich., Oct. 23, 1864.

Heogler, Joseph. A member of company K, First Sharpshooters. Died Oct. 24, 1864. Buried in Elmwood Cemetery, Detroit, Mich.

Kadah, Joseph – PRIME NAME, from Adjutant General Record
Heogler, Joseph – Adjutant General Record, second entry
Hagler, Joseph – Name on gravestone in Detroit Elmwood Cemetery
Haegler, Joseph – Name on Elmwood Cemetery burial records

RESIDENCE IN 1863: Elbridge township, Oceana County S-AGR

DEATH DATE: October 23, 1864 *(Died at St. Mary's Hospital, Detroit)*

GRAVE: Wayne Co., Detroit / Elmwood Cemetery, Section S (Military stone)
Note: Section S is a Soldiers Section. There is a flagpole in front of it, making it easy to find. Joseph's grave is in the right half, in the second row.
Two stones mark his grave:
1. New gravestone says Joseph Hagler, Pvt, Co. K, 1st Mich Sharp Shooters
2. Older military stone says Jos. Kaichter, 1st Co., 27th Mich. Infantry

PENSION FILE: No Pension File exists
VETERAN CENSUS: 1888 No / 1890 No / 1894 No
RAY HEREK's BOOK: Pages: 417, 434(Roster)

SPECIAL INFORMATION:
One of the mystery soldiers of Company K is Joseph Heogler / Hagler. Where did he come from?
In my opinion, **Joseph Heogler = Joseph Kadah** for the following reasons:
The first names match. One man died in a Detroit hospital on Oct 23, the other man appeared in a Detroit cemetery the following day. I'm sure these two men are the same person.
Both the Adjutant General record and Elmwood Cemetery records agree, he died Oct 23, 1864.
The Adjutant General record for Joseph Heogler mistakenly says he died Oct 24 – that was actually his burial date. S-burial book in the office of Elmwood Cemetery, Book E, pg. 25
There is a Veterans Monument in Hart, Michigan (Oceana County) with his name on it.

NOTES FOR HISTORIANS:
The original gravestone for Joseph Kadah in Elmwood cemetery was a military stone that said "Jos. Kaichter, 1st Co., 27th Mich. Inf." (=Perrin's Sharpshooters).

JOSEPH KAKAKEE 061

Ka-ka-kee, Joseph. Enlisted in company K, First Sharpshooters, Aug. 21, 1864, at Grand Rapids, for 1 year, age 35. Mustered Sept. 1, 1864. Discharged at Alexandria, Va.

Kakakee, Joseph – PRIME NAME, from Pension File
Me-she-ka-kack, Joseph – Pension File, aka name
Me-she-ka-kuch, Joseph – State Death Record name

RESIDENCE IN 1864: Chester, Ottawa County S-Descriptive Roll, State Archives
1857: Ontario, Canada, at the Cape Croker Indian Reserve near Owen Sound
1865 – 1901: Mason County, Custer twp. & Branch twp. S-pf-061 and Land Records.
DEATH DATE: May 12, 1901 S-State Death Record
GRAVE: Mason County, Custer / Custer Riverside Cemetery (Military stone)
(Location: Custer is east of Ludington on US-10. The cemetery is south of town on Custer road. Look in the back of the cemetery for a dirt road leading through the old Indian Burial Ground. His grave is on a high bluff overlooking the Pere Marquette river. Every summer, countless canoes will pass within a hundred yards of his gravestone)
PENSION FILE: Yes, see Appendix A; 1883 Pension Roll: No
VETERAN CENSUS: 1888 No / 1890 No / 1894 Yes
RAY HEREK's BOOK: Pages: 434(Roster)

SPECIAL INFORMATION:
He enlisted in the army as "Kakakee". The name means Pigeon Hawk. S-ccn061a
For most of his life, he used the name "Joseph Me-she-ka-kack". S-ccn061b
He was born in Michigan, probably as a member of one of the Grand River bands. Went to Canada and lived many years with the Chippewa on the Cape Croker Indian Reserve (renamed Nawash in the 1980's) on the Bruce Peninsula, Georgian Bay, Ontario. There he married Mary Sunego.
Two known children: Martha b. 1846, and Moses b. 1850. S-pf-061 (pension file for Joseph Kakakee)
He signed a treaty in1857. This indicates he was an important and respected member of the tribe. Then, for unknown reasons, he left his family and returned to Michigan. Joseph Kakakee enlisted in Company K at the age of 35 and saw action in Virginia at the Battle of Fort Steadman, and the charge against the lines at Petersburg on April 2, 1865. He was never wounded, but claimed to have injured his back while dismantling enemy breastworks. His back injury made him an invalid for the rest of his life.
After the war he lived in Michigan in Mason County.
He was the uncle of Company K soldier Aaron Sahgahnahquato. S-pf-061 (Joseph Kakakee pension file)

NOTES FOR HISTORIANS:
The date and place of his discharge is a problem. His official army record says "Discharged at Alexandria, Va.", no date given. Pension file says he was discharged June 3, 1865 at Detroit, MI. The Adjutant General Annual Report spells the name Joseph Kukakee and says he was from Macomb County In my opinion, Macomb is a clerical error, and it should say Mason County.

SAMUEL KAQUATCH 062

Kaquatch, Samuel, Little Traverse. Enlisted in company K, First Sharpshooters, June 15, 1863, at Little Traverse, for 3 years, age 18. Mustered July 20, 1863. Wounded in action June 17, 1864. Killed by car in Philadelphia, Pa., Aug. 2, 1864. Buried in National Cemetery at Philadelphia, Pa. Grave No. 614.

Kaquatch, Samuel – PRIME NAME - Adjutant General Record
Kagwaich, Samuel – Pension File name
Kagwaitch, Samuel – Gravestone name

RESIDENCE IN 1863: Little Traverse (=Harbor Springs), Emmet County S-AGR
DEATH DATE: August 2, 1864 S-Adjutant General Record, and Death Certificate
Died in a train accident in Philadelphia, PA S-pf-062 (pension file for Samuel Kaquatch)
GRAVE: Philadelphia National Cemetery, Philadelphia, PA. Grave # 614

PENSION FILE: Yes, see Appendix A; 1883 Pension Roll: No
VETERAN CENSUS: 1888 No / 1890 No / 1894 No
RAY HEREK's BOOK: Pages: 182, 396, 420, 434(Roster)

SPECIAL INFORMATION:
The only way his mother could get a pension was by an ACT OF CONGRESS !
When Samuel Kaquatch was wounded in the right arm on June 17, 1864, he was sent to a hospital in Philadelphia to recover from his injury. Six weeks later he died in a tragic accident. According to official records, he received a pass to leave the hospital and walk around the city for exercise. During this walk he was run over by a train and killed. The inquest determined he was intoxicated. That ends the story of Samuel Kaquatch. But for his parents back home in Harbor Springs, it was the beginning of a 25 year fight with the government for pension benefits. The government decided Sam did not "die while in the line of duty." So the pension application was Rejected. The mother, Sophia Kaquatch, refused to accept that answer. She applied over and over again. Many years later, in 1893, the government changed its mind. To cut red tape and speed her claim, a special ACT OF CONGRESS was passed, granting her a pension. They reasoned, the fact Sam was wounded in battle outweighed the one mistake he made while on pass from the hospital. His mother, Sophia, died March 29, 1903. She had outlived her husband and all her children. This brave Odawa woman is buried in an unmarked grave in Lakeview Cemetery, Harbor Springs.

NOTES TO HISTORIANS:
Correct death date is August 2, 1864, according to 3 sources: Philadelphia City Death Certificate (there is a copy in his pension file), the Railroad Accident Log Book, and the newspaper story in the Philadelphia Ledger (see the Philadelphia Ledger, Thursday, August 4, 1864, Local Affairs).
His gravestone says he died Oct. 1, 1864. That is an error. First he was buried in Glenwood; later he was reburied in the National Cemetery in Philadelphia. October1 was the reburial date.
The ACT OF CONGRESS mentioned above was H.R. 8498, passed February-March 1893.

THOMAS KECHITTIGO 063

Chetego, Thomas K., East Saginaw. Enlisted in company K, First Sharpshooters, as Sergeant, May 3, 1863, at Buena Vista, for 3 years, age 27. Mustered May 26, 1863. Mustered out at Delaney House, D. C., July 28, 1865. Present residence; Grayling, Mich.

Kechittigo, Thomas – PRIME NAME, from Pension File
Chetego, Thomas K. – Adjutant General Record name
Ne-o-de-geshik – Pension File says this was his Anishinabe name

RESIDENCE IN 1863: Buena Vista township, Saginaw County S-AGR
In 1890: Pinconning, Bay County S-1890 Veteran Census
In 1894: Grayling, Crawford County S-1894 Veteran Census

DEATH DATE: April 24, 1916 S-pf-063
GRAVE: Crawford County, Grayling / Elmwood Cemetery – Block 3, Lot 36
(Note: *His unusual gravestone looks like a tree trunk made from stone. It was provided by a fraternal order called the Woodmen of America.)* Cemetery Location: To find Elmwood Cemetery, take I-75 to Exit 256, this will exit on North Down River Road (Hwy F-32). Go west on North Down River Road till you reach McClellan Road, also known as Business I-75. Elmwood cemetery is on the north-east corner of this intersection.
PENSION FILE: Yes, see Appendix A; 1883 Pension Roll: No
VETERAN CENSUS: 1888 Yes / 1890 Yes / 1894 Yes
RAY HEREK's BOOK: Pages: 24, 34, 119, 148, 352, 377, 378, 390, 434(Roster), 469

SPECIAL INFORMATION:
His pension file specifically says "Kechittigo" is the correct spelling of his name. He was an extremely strong and large man, and when he enlisted they immediately made him a Sergeant (S-Herek pg 34). At the Battle of Spotsylvania he had a close call with an artillery shell (his left forearm was fractured S-Herek pg 390). **At the hospital he was photographed standing under a tree. The now-famous photograph is titled "Wounded Indians from the Wilderness on Marye's Heights".** His date of birth is uncertain - pension file says May 15, 1835, but his gravestone says May 3, 1836. The newspaper in Grayling is called the "Crawford Avalanche". It ran a long and well researched story about Thomas Kechittigo in a special issue called "Historical Keepsake Edition" Thursday, August 21, 2003. Author was Fay Bovee. The Clarke Library in Mt. Pleasant has a copy. Anyone researching this soldier should start there.
In 1886 there was a reunion of the 1st Mich. SS in Jackson. Tom attended this reunion (S-ccn063a).
In 1884, Thomas Kechittigo joined the G.A.R., a veterans organization like the VFW of today. He was a charter member of Scully Post No. 265 in Omer, Algonac Co. When Tom moved to Grayling in Crawford County, he joined the Marvin G.A.R. Post No. 240 as member #103. One newspaper story says he was the Post's color bearer (meaning he held the flag when they marched in parade) S-ccn063b. **There is a photograph taken in 1915, the year before he died, of Tom dressed up in his best clothes and proudly wearing his G.A.R. badge (**S-ccn063c).

JOHN KEESES 064

Keesses, John. Enlisted in company K, First Sharpshooters, Aug. 23, 1864, at Kalamazoo, for 2 years, age 30. Mustered Aug. 23, 1864. Substitute for Edward W. Gibson. Mustered out at Delaney House, D. C., July 28, 1865.

Keeses, John – PRIME NAME, Adjutant General Record
Keesis, John – Descriptive Roll, State Archives
Kesis, Jno – Gravestone name
Kesses, John – Pension File name

RESIDENCE IN 1864: Richland, Kalamazoo County S-Descriptive Roll, State Archives
In 1867: Wayland, Allegan County (place of death) S-pf-064

DEATH DATE: July 15, 1867 (*Died near Wayland, Allegan County*) S-pf-064

GRAVE: Allegan Co., Bradley / Bradley Indian Mission Cemetery (Military stone)

PENSION FILE: Yes, see Appendix A; 1883 Pension Roll: No

VETERAN CENSUS: 1888 No / 1890 No / 1894 No

RAY HEREK's BOOK: Pages: 434(Roster)

SPECIAL INFORMATION:
He was paid to be a substitute for Edward W. Gibson, a man who was going to be drafted. Judging from the fact that John Keeses lived near Wayland in Allegan County, and also from the fact that he is buried in the Bradley Indian Mission Church cemetery, he probably belonged to the Match-E-Be-Nash-She-Wish Band of Pottawatomi (aka Gun Lake band). But this is not a proven fact – further research needed.
Shortly before the war, John Keeses married Mary Nottawa. When he enlisted in the army at the age of 30, he left behind his wife Mary and their three year old baby daughter, Eliza. S-pf-064
John Keeses survived the war, but died of consumption (tuberculosis) at the Bradley Indian Mission on July 15, 1867, according to a statement by fellow Company K soldier Thomas Wesaw. S-pf-064
His only surviving child was Eliza. She married Adam Shipman, an Indian from Walpole Island, on February 28, 1883. The couple lived in the Athens, Michigan area for the next 30 years. S-pf-064

NOTE TO HISTORIANS:
Disregard the note at the bottom of his Pension File Index Card saying he died October 20, 1867. This is a clerical error. The date October 20 is the birthdate of his daughter Eliza, and the other documents in his pension file indicate that July 15, 1867 is the correct death date for John Keeses.

SIMON KEJIKOWE 065

Keji-kowe, Simon, Burt. Enlisted in company K, First Sharpshooters, July 6, 1863, at Little Traverse, for 3 years, age 23. Mustered July 20, 1863. Absent wounded Sept., 1864. No further record.

Keji-kowe, Simon – PRIME NAME, Adjutant General Name
Gijigowi, Simon – Genealogy records S-ccn065a
Kigegoway, Simon – Genealogy records S-ccn065a

RESIDENCE IN 1863: Burt (=Burt Lake), Cheboygan County S-AGR

DEATH DATE: Unknown

GRAVE: Location Unknown

PENSION FILE: No Pension File exists

VETERAN CENSUS: 1888 No / 1890 No / 1894 No

RAY HEREK's BOOK: Pages: 148, 391, 404, 435(Roster)

SPECIAL INFORMATION:
Simon Kejikowe was a member of the Burt Lake band.
What's more, he was the son of the band's leader, Chief Joseph Gijigowe. S-Shier pg 207
In the summer of 1863, word traveled through the tribes that Lt. Garrett Graveraet was calling for Anishinabe volunteers to join the army. Simon answered the call and went to Harbor Springs to enlist. One year later he was wounded in action at Spotsylvania, VA. Simon was one of thousands sent to makeshift army hospitals, where the staff was too busy to write down names of the dying and the dead. So there is no record of when he died, or where he was buried. A good guess for the location of his grave would be Fredericksburg National Cemetery, where the mortally wounded men from Spotsylvania were buried in plots marked "Unknown Soldier".

NOTE TO HISTORIANS:
In the records of the 1st Michigan Sharpshooters is a puzzling casualty report for the action at Peebles Farm on September 30, 1864. In this report Simon Kejikowe is listed as one of the wounded. The man should have been long dead by then, so why is he on the report? Best guess, this is a simple error in the records. S-Herek pgs 391, and 404
Simon Kejikowe was one of three men from Burt Lake who enlisted into Company K during the Civil War. The other men were Simon Sanequaby and Charles Porcellay.

AMABLE KETCHEBATIS 066

Kitcherbalist, Amable, LaCroix. Enlisted in company K, First Sharpshooters, Aug. 24, 1863, at Northport, for 3 years, age 22. Mustered Oct. 4, 1863. Taken prisoner July 30, 1864. No further record.

Chibadice, A. A member of Company K, First Sharpshooters. Missing in action at Petersburg, July 30, 1864.

Ketchebatis, Amable – PRIME NAME, from Pension File
Kitcherbalist, Amable – Adjutant General Record
"Amable John Baptist" – His correct name, from Pension File
"Amabilis Egawanige" – His name, from church record in Pension File
Ketchavatis, Elizabeth – 1883 Pension Roll (Widow's name)
Ketchabaptist, Amab – 1890 Veteran Census (entry for his widow)
Chibadice, A. – Adjutant General Record (see note below)

RESIDENCE IN 1849: Manitoulin Island, Canada (married there) S-pf-066
Residence in 1863: LaCroix (=Cross Village), Emmet County S-AGR
DEATH DATE: February 22, 1865 S-Shier pg 218
GRAVE: Danville National Cemetery in Danville, Va., Grave #1167 S-Shier pg 219
(*Gravestone #1167 says "J. Amos, MICH"*) S-Shier pg 219
PENSION FILE: Yes, see Appendix A
1883 Pension Roll: Yes, Emmet County - Cross Village (Widow)
VETERAN CENSUS: 1888 No / 1890 Yes (widow) / 1894 No
RAY HEREK's BOOK: Pages: 403, 435(Roster)

SPECIAL INFORMATION: He and his wife Lizette (aka Josette) came from Manitoulin Island, Canada, and settled in Cross Village. His true name was Amable John Baptist, but since he was a big man, everyone called him Kitchi-Batise (Big Baptist). Army records say he was 22 when he enlisted, but that is a clerical error. He was actually 32 (S-pf-066, church-record says born in 1830). Personally recruited for Company K by Garrett Graveraet. Captured during the tragic Battle of the Crater on July 30, 1864. His wife received a letter from him in August, saying he was all right. She never heard from him again. After the war, witnesses said he died around March 1, 1865 at the prison in Danville, VA. He left behind a widow and four children: Margaret, Catherine, Mary and Joseph. Lizette received a pension from the government for 53 years. She died September 7, 1918, and is buried in Cross Village. Her State Death Record has the name "Elisabeth Tchepates".

NOTES FOR HISTORIANS: The 1890 Veteran Census says this soldier died at Andersonville. Not true, he was sent to a prison in Danville, VA, and stayed there till his death in February of '65. **Amable Ketchebatis and A. Chibadice are the same person.** His pension file proves this fact. His daughter, Margaret, sometimes used Chibadice as her last name.

BENJAMIN KEWACONDA 067

Kewaconda, Benjamin. Enlisted in company K, First Sharpshooters, Feb. 24, 1864, at Mackinaw, for 3 years, age 21. Mustered April 29, 1864. Mustered out at Delaney House, D. C., July 28, 1865.

Kewaconda, Benjamin – PRIME NAME, Adjutant General Record
Kewacondo, Benjamin – Enlistment Paper

RESIDENCE IN 1864: St. Ignace, Mackinac County S-Descriptive Roll, State Archives

DEATH DATE: December 17, 1886 S-Ancestry.com / U.S. Civil War Soldier Records and Profiles
GRAVE: Roanoke, Virginia S-Ancestry.com / U.S. Civil War Soldier Records and Profiles

PENSION FILE: No Pension File exists
VETERAN CENSUS: 1888 No / 1890 No / 1894 No
RAY HEREK's BOOK: 435(Roster)

SPECIAL INFORMATION:
Born in Georgian Bay, Ontario, Canada. S- Army Enlistment Records
He probably belonged to the Bear River Band in Emmet County. S-ccn067a
His home in 1864 was Point St. Ignace on the north side of the Straits of Mackinac. (s-DR)
Enlisted on Mackinac Island Feb. 29, 1864 (not Feb. 24, see below). Occupation: Boatman.
Three men from St. Ignace enlisted in Company K on Mackinac Island during February and March of 1864. Their names were Kewaconda, Hamlin, and Jeandron. All three were mustered on April 29, 1864. Two were captured and perished at Andersonville. Kewaconda survived the war.

ERROR IN THE RECORDS: Adjutant General Record says he enlisted February 24, 1864. This is wrong. His enlistment paper clearly shows the correct enlistment date is February 29, 1864.

According to the internet database called U.S. Civil War Soldier Records and Profiles, this man died on December 17, 1886 and is buried in Roanoke, Virginia. All information found on the internet needs to be double checked for accuracy, so further research is needed to confirm this.

After the Civil War, Benjamin Kewaconda joined the army again! His new regiment was the U.S. 10th Cavalry (famous as a "Buffalo Soldier" unit). He was a trooper in Company E. Enlisted in Detroit in 1867. Stationed at Fort Sill, Oklahoma. Further research needed.

NOTES FOR HISTORIANS:
The exact date and place when Benjamin Kewaconda entered the ranks of Company K is uncertain. Army records are vague and contradictory. It is likely he was mustered into the army in Virginia on the day he arrived at the camp of the 1st Michigan Sharpshooters. That would be April 29, 1864, during a six day halt at Warrenton Junction. But this is not a proven fact – further research needed.

JOHN KINEWAHWANIPI 068

Kiniewahasoipi, John. Enlisted in company K, First Sharpshooters, Feb. 14, 1865, at Grand Rapids, for 1 year, age 20. Mustered Feb. 14, 1865. Died of disease March 5, 1865. Buried in Poplar Grove National Cemetery, Petersburg, Va.

Kinewahwanipi, John – PRIME NAME, from Pension File
Kiniewahasoipi, John – Adjutant General Record
Kenewangic, John – Adjutant General Annual Report
Kenewancic, Jno – Name on gravestone
Kennewanche, John – Name on Poplar Grove Cemetery records
Ke-new-nuh-ne-ba, John – Pension File, aka name
Penegwah, John – Pension File, aka name
Pashabe, John – Diary of Rev. Smith – March 26, 1865

RESIDENCE IN 1865: Leelanau twp., Leelanau Co. S-Descriptive Roll, State Archives
DEATH DATE: March 3, 1865 (*Died in an army hospital*) S-pf-068
For the correct date of death, see note below
GRAVE: Poplar Grove National Cemetery, near Petersburg, VA. Grave # 0189
Cemetery records list him as "John Kennewanche", but gravestone says "Jno Kenewancic"

PENSION FILE: Yes, under three names: Kinewahwanipi, Penegwah, and Ke-new-nun-ne-ba (see Appendix A) 1883 Pension Roll: No
VETERAN CENSUS: 1888 No / 1890 No / 1894 No
RAY HEREK's BOOK: Pages: 290, 417, 435(Roster)

SPECIAL INFORMATION:
His home was on the Leelanau peninsula, near Northport. S-Descriptive Roll, State Archives
Born May 16, 1848 in Ahgosa Town, Leelanau Co. Occupation: Saw Mill worker. S-Jim Lalone
While Payson Wolf was home on furlough, recovering from his stay at Andersonville, Payson persuaded John Kinewahwanipi to enlist in Company K. S-Herek pg 290
Three weeks after he enlisted, poor John died from double pneumonia. S-pf-068
Mentioned in the diary of the missionary in Northport, the Rev. Smith, as "John Pashabe":
"[one] of the young Indians who went with Payson died after three days sickness – a letter brings the news – John Pashabe." S-Diary of the Rev. George N. Smith, entry for March 26, 1865
There are three separate pension files for this soldier, under three names:
1 Kinewahwanipi, John
2 Ke-new-nuh-ne-ba, John
3 Penegwah, John
ERROR IN THE RECORDS: Adjutant General Record says he died March 5, 1865.
This is wrong. Correct death date is March 3, 1865. S-ccn068a

THADDEUS LAMOURANDERE 069

Lamourandere, Thaddeus, Pentwater. Enlisted in company K, First Sharpshooters, July 4, 1863, at Elbridge, for 3 years, age 21. Mustered July 11, 1863. Killed in action at Spottsylvania, Va., May 12, 1864.

Lamourandere, Thaddeus – PRIME NAME, Adjutant General Record
Lamourandiex, Thaddeus – Herek's book, page 435
Lamorander, Taddy – Pension File, aka name

RESIDENCE IN 1863: Elbridge township, Oceana County S-AGR

DEATH DATE: May 12, 1864 (*Killed in action at Spotsylvania, VA*)

GRAVE: Location Unknown

PENSION FILE: Yes, see Appendix A
1883 Pension Roll: No

VETERAN CENSUS: 1888 No / 1890 No / 1894 No

RAY HEREK's BOOK: Pages: 389, 435(Roster)

SPECIAL INFORMATION:
Member of the Grand River Band of Ottawa.
He was one of the 25 men who enlisted on the 4th of July at Elbridge. S-Herek pg 35
Killed in action at the Battle of Spotsylvania on May 12, 1864.
There is a Veterans Monument in Hart, Michigan (Oceana County) with his name on it.

NOTE TO HISTORIANS:
The Adjutant General Annual Report says his residence was "Ionia County". This is a clerical error, should read "Oceana County".

DAVID LIDGER 070

Lidyer, or Lidger, David. Enlisted in company K, First Sharpshooters, Feb. 4, 1865, at Detroit, for 3 years, age 28. Mustered Feb. 4, 1865. Mustered out at Delaney House, D. C., July 28, 1865.

Lidger, David – PRIME NAME, Adjutant General Record
Lidyer, David – Adjutant General Record, aka name
Lidger, Daniel – Ray Herek's Roster (Note: Daniel) S-Herek pg 496
Lidger, Daniel – Adjutant General Annual Report (note: Daniel)
Leidyer, David – Descriptive Roll, State Archives

RESIDENCE IN 1865: Mendon, St. Joseph County S-Descriptive Roll, State Archives

DEATH DATE: Unknown

GRAVE: Location Unknown

PENSION FILE: No Pension File exists

VETERAN CENSUS: 1888 No / 1890 No / 1894 No

RAY HEREK's BOOK: Pages: 435(Roster), 496

SPECIAL INFORMATION
David Lidger was one of two Iroquois men in Company K. A roster in State Archives of Michigan says he was born in Oneida, New York, same as Joseph Hannin. These two men had a lot in common – both enlisted at the same time with the same recruiter, and were credited to Mendon, MI in St. Joseph County. We know for certain that Joseph Hannin was an Iroquois born at Oneida Castle, New York. Around 1840 his tribe sold their land to settlers and moved as a group to Canada. Nowdays this tribe is officially called the Oneida Nations of the Thames, living on Oneida Reserve 41, aka The Oneida Settlement. It is very likely that David Lidger was also an Iroquois of the Oneida Nations, but this is not a proven fact. Further research is needed.
Most records say his first name was "David". However, historian Raymond Herek and the Adjutant General Annual Report says his first name was "Daniel". More research is needed.
The Adjutant General Record gives two possible last names: Lidger, and Lidyer.
It is a common clerical error to write the letter "g" with the top of the letter open, so it looks Like the letter "y". For this reason I use "Lidger" as the PRIME NAME.
Note that historian Raymond Herek also prefers the "Lidger" spelling.
David Lidger survived the war and was discharged from the army.
Nothing is known about his life after the war.
Source of birth information: Archives of State of Michigan, record 5914_Ovs_70 folder 03 doc 37

JOSIAH LIGHT 071

Light, Josiah, Pentwater. Enlisted in company K, First Sharpshooters, July 4, 1863, at Elbridge, for 3 years, age 18. Mustered July 11, 1863. Promoted Corporal March, 1865. Mustered out at Delaney House, D. C., July 28, 1865.

Light, Josiah – PRIME NAME – all sources agree on spelling
Light, Joseph – See note below

RESIDENCE IN 1863: Elbridge township, Oceana County S-AGR

DEATH DATE: October 1884 (hunting accident – See Ludington Record, October 23, 1884)

GRAVE: Mason County, Eden township / Eden Indian Cemetery S-Shier pg 226

PENSION FILE: No Pension File exists

VETERAN CENSUS: 1888 No / 1890 No / 1894 No

RAY HEREK's BOOK: Pg. 435(Roster)

SPECIAL INFORMATION:
He was one of those 25 men who volunteered on the 4th of July in Elbridge.
Promoted to Corporal. Seems to have gone through the war without illness or injury.
Definitely was at the Battle of the Crater. S-pf-131 (pension file for John Wabesis)
Mustered out with the rest of Company K at the end of the war.
After the war, he married and lived on a farm in Oceana County. S-1870 US Census
Later he moved to Mason County, Eden township, and worked as a laborer. S-1880 US Census
His life had its share of tragedy. First he lost a leg at the knee. Then some land he claimed as a homestead was unfairly taken from him. In 1883 he lost his wife and two children. Finally, in October of 1884, he died in a tragic hunting accident. He was buried in the Indian Cemetery in Eden township, Mason County, near his wife and children. *This story was discovered and first published by Quita V. Shier in her book "Warriors In Mr. Lincoln's Army" on page 226. See also the Ludington Record, October 23, 1884, page 8, col 1, for details of the accident.*

NOTE TO HISTORIANS:
An Indian named **Joseph Light**, described as "a soldier in the war of the rebellion who lost a leg in the service" turns up in 1875, complaining that he had been unfairly removed from his homestead (S-ccn071a). According to the news report of the hunting accident, Josiah Light had lost a leg at the knee. So it appears the "Joseph Light" in the news story is same as "Josiah Light" of Company K. However, he did not loose his leg in the war, but sometime after his army service.

LOUIS MARKS 072

Marks, Louis, Bear River. Enlisted in company K, First Sharpshooters, June 1, 1863, at Bear River, for 3 years, age 24. Mustered June 22, 1863. Taken prisoner at Petersburg, Va., June 17, 1864. Confined at Andersonville, Ga. Released at Savannah, Ga., Nov. 20, 1864. Discharged at Annapolis, Md., June 28, 1865.

Marks, Louis – PRIME NAME, most sources agree with this spelling
Mark, Louis – Name in the Charles Campbell diary (see below)
O-Taw-Naw-Zhe, Louis – Durant Roll, Field Notes #50-pg29

RESIDENCE IN 1863: Bear River (=Petoskey), Emmet County S-AGR
DEATH DATE: April 2, 1877 S-pf-072 (pension file for Louis Marks)
GRAVE: Location Unknown (Probably in Emmet County., Bear Creek township)
PENSION FILE: Yes, see Appendix A; 1883 Pension Roll: No
VETERAN CENSUS: 1888 No / 1890 No / 1894 No
RAY HEREK's BOOK: Pages: 398, 435(Roster), 495

SPECIAL INFORMATION:
His Anishinabe name was Louis O-Taw-Naw-Zhe. S-Durant Roll Field Notes #50-pg29
He was one of the 15 soldiers from Company K captured at Petersburg, VA, on June 17, 1864, and sent to the infamous Andersonville Prison in Georgia.
Before Louis Marks left Petersburg as a POW, a resident of that city named Charles Campbell heard about the Indians who had been captured and went to see them. He recorded the meeting in his diary. Louis Marks is mentioned by name as one of the men he talked to.
For the details of this remarkable event, see the unpublished diary of Charles Campbell, entry for June 18, 1864. This diary can be found at the College of William and Mary, Williamsburg, VA, in the Swem Library Special Collections, in the Charles Campbell Papers, catalog heading Charles Campbell's diaries (1861-1864).
Louis Marks was a POW at Andersonville for five months. He survived the ordeal and made it home to Michigan.
Received extra 3 months' pay for being a POW. S-Muster Out Records, State Archives, Lansing, Reel #11629.

NOTES FOR HISTORIANS:
When Louis came back from the war his father, Enos Marks of Petoskey, was over 70 years old. Louis took care of him and was his sole means of support. Tragically, Louis died in 1877. His father was destitute and applied to the government for a pension in 1880. The claim was rejected.
The building in Petersburg, VA., where Louis Marks was imprisoned and visited by Charles Campbell is today marked with a historical plaque mentioning "captured Native American soldiers from Michigan". It is on the corner of North Market Street and Grove Avenue. Local people call it the "Peter Jones Trading Station".

FRANK MARQUETTE 073

Marquette, Frank. Enlisted in company K, First Sharpshooters, Oct. 24, 1864, at Grand Rapids, for 1 year, age 35. Mustered Oct. 24, 1864. Joined regiment at Peebles' Farm, Va., Nov. 4, 1864. Discharged at Washington, D. C., June 27, 1865.

Marquette, Frank – PRIME NAME, Adjutant General Record
Marquette, Francis – Descriptive Roll, State Archives

RESIDENCE IN 1864: Assyria, Barry County S-Descriptive Roll, State Archives

DEATH DATE: Unknown

GRAVE: Location Unknown

PENSION FILE: No Pension File exists

VETERAN CENSUS: 1888 No / 1890 No / 1894 No

RAY HEREK's BOOK: Pages: 273, 435(Roster)

SPECIAL INFORMATION:
According to the Descriptive Roll in State Archives, his occupation before the war was "hunter", and he was from Assyria, in Barry County.
There is no tribe or reservation at Assyria, so why he was living there is a puzzle.
Nearest reservation is 30 miles away, the Nottawaseppi Huron Reservation located south of Battle Creek.

NOTE TO HISTORIANS:
Most of Company K was mustered out on July 28. Notice that Frank Marquette is discharged June 27. To me, this suggests he was discharged from a hospital. And that means he was either seriously wounded or seriously ill. Further research needed.

JAMES MASHKAW 074

Mash-kaw, James, Pentwater. Enlisted in company K, First Sharpshooters, July 4, 1863, at Pentwater, for 3 years, age 31. Mustered July 11, 1863. Killed in action at Spottsylvania, Va., May 12, 1864.

Mash-kaw, James – PRIME NAME, Adjutant General Name
Kah-kuh-ge-wa – Marriage record, aka name.
Kaw-gaw-ge-way – Name on Probate Record, Oceana County 1871
Mash Kay, Chic-ah-mil-qua – 1883 Pension Roll (widow)

RESIDENCE IN 1863: Pentwater, Oceana County S-AGR

DEATH DATE: May 12, 1864 (*Killed in action at Spotsylvania, VA*)
GRAVE: Location Unknown

PENSION FILE: Yes, see Appendix A
1883 Pension Roll: Yes, Emmet County - Cross Village (Widow)

VETERAN CENSUS: 1888 No / 1890 No / 1894 No
RAY HEREK's BOOK: Pages: 148, 343, 346, 389, 435(Roster)

SPECIAL INFORMATION:
James and John Mashkaw were brothers who joined the army together.
They enlisted on the 4th of July, fought side by side in battle, and they died together in the fierce fighting near Spotsylvania Court House on May 12, 1864.
There is a Veterans Monument in Hart, Michigan (Oceana County) with the names of the Mashkaw brothers on it.
This soldier had two names: James Mashkaw was his official name, but to his family and friends, he was known as Kaw-gaw-ge-way. S-Probate Record, Oceana County 1871
James Mashkaw was married August 6, 1858 by a Justice of the Peace in Pentwater, MI. Marriage record says his name was James Mash-Kaw, *aka* Kah-kuh-ge-wa, and he was age 28. He married Keh-guh-me-qua, age 37. There were no children. S-pf-074 (James Mashkaw pension file)
James was killed in action at the battle of Spotsylvania. His widow applied for a pension as "Mrs. Chig-ah-me-gua Mash-Kaw". It was approved, and she collected $8.00 a month for the rest of her life. In 1883 she was living in Cross Village (Emmet County).

NOTES FOR HISTORIANS:
They called him "Jim", and he was the son of a Chief.
His father was John Maishkaw Sr, Chief of the band. His mother was Wahonoquay.
His grandfather signed the 1821 Treaty of Chicago, as Chief Kee-o-to-aw-be, also known as Kewaytowaby. S-Shier pg 232.

JOHN MASHKAW 075

Mash-kaw, John, Pentwater. Enlisted in company K, First Sharpshooters, July 4, 1863, at Elbridge, for 3 years, age 22. Mustered July 11, 1863. Killed in action at Spottsylvania, Va., May 12, 1864.

Mash-kaw, John – PRIME NAME, Adjutant General Record
Mashkan, John – Adjutant General Annual Report
Mash-kah, John – Pension File name
Marcum, John – Descriptive Roll, State Archives, aka name
Wash kah, Bavrg-man – 1883 Pension Roll Name (Father)

RESIDENCE IN 1863: Pentwater, Oceana County S-AGR

DEATH DATE: May 12, 1864 (*Killed in action at Spotsylvania, VA*)

GRAVE: Location Unknown

PENSION FILE: Yes, see Appendix A
1883 Pension Roll: Yes, Mason County – Custer (Father)

VETERAN CENSUS: 1888 No / 1890 No / 1894 No

RAY HEREK's BOOK: Pages: 148, 343, 346, 389, 435(Roster)

SPECIAL INFORMATION:
James and John Mashkaw were brothers who joined the army together.
They enlisted on the 4th of July, fought side by side in battle, and they died together in the fierce fighting near Spotsylvania Court House.
There is a Veterans Monument in Hart, Michigan (Oceana County) with the names of the Mashkaw brothers on it.
After the war, the parents of James and John Mashkaw received a government pension and lived their final years in Custer township, Mason County.
His father was John Maishkaw Sr, Chief of the band. His mother was Wahonoquay.
His grandfather signed the 1821 Treaty of Chicago, as Chief Kee-o-to-aw-be, also known as Kewaytowaby. S-Shier pg 232.

 All Rights Reserved by Chris Czopek

THOMAS MILLER 076

Miller, Thomas, Omena. Enlisted in company K, First Sharpshooters, June 13, 1863, at, Northport, for 3 years, age 19. Mustered June 22, 1863. Deserted at Philadelphia, Pa., June 6, 1865.

Miller, Thomas – PRIME NAME, Adjutant General Record
Kah-gee, Thomas – His native name S-pf-076 (pension file for Thomas Miller)

RESIDENCE IN 1863: Omena, Leelanau County S-AGR
DEATH DATE: October 15, 1864 S-pf-076 (pension file for Thomas Miller)
(*Died of disease at his home in Omena, Michigan, while on convalescent leave*)
GRAVE: Leelanau County, Omena / Omena Presbyterian Church Cemetery
Look for a Military gravestone in the graveyard behind the big, white, wood church
PENSION FILE: Yes, see Appendix A; 1883 Pension Roll: No
VETERAN CENSUS: 1888 No / 1890 No / 1894 No
RAY HEREK's BOOK: Pages: 435(Roster)

SPECIAL INFORMATION:
Thomas Miller grew up in the Grand Traverse Bay region. His native name was Thomas Kah-gee. His parents were married at Old Mission by the Rev. Dougherty. He was their only child. When the Rev. Dougherty moved to New Mission (Omena), the family moved with him to the Leelanau peninsula. The big white Presbyterian church in Omena was their church. Thomas is buried in the graveyard behind this church. His elderly father had two names: Thomas K. Miller, and Henry Kah-gee. The father lived in Charlevoix in the 1870's with his second wife and worked there as a fisherman. He applied for a pension from the government in 1870, and this pension file is the main source of what know about the soldier Thomas Miller and his family.

NOTES FOR HISTORIANS:
His official army record (top of this page) says he was a deserter. Not true! What happened is this: Before his first battle, Thomas Miller became seriously ill. When the regiment was passing through Annapolis, MD, in April 1864, he was left behind in a hospital. He did not improve.
He was transferred to one of the finest hospitals in the Civil War, Mower Hospital in Philadelphia. Still, he did not improve. The hospital gave him 20 days leave to go home to Michigan and recover. His condition (lung disease) became worse, and he died while home on convalescent leave. In the confusion of war, neither Mower Hospital nor his regiment in Virginia received news of his death. That is why he was wrongly accused of being a deserter.
His father's name on the pension file index card is spelled Kahgel. This is a clerical error, correct spelling is Kahgee. The Rev. Amos Kahgee, well known Northport Indian minister, was no relation. Thomas Miller, the soldier, was never married and has no descendants.

EDWARD MISISAIUS 077

Misisaius, Edward, LaCroix. Enlisted in company K, First Sharpshooters, June 14, 1863, at Northport, for 3 years, age 25. Mustered Oct. 4, 1863. Absent wounded. No further record.

Misisaius, Edward – PRIME NAME, Adjutant General Record
Misisins, Edward – Descriptive Roll, State Archives
Minissisance, Edward – Descriptive Roll aka name

RESIDENCE IN 1863: LaCroix (=Cross Village), Emmet County S-AGR
(Enlisted in Northport, Leelanau County)

DEATH DATE: Unknown

GRAVE: Location Unknown

PENSION FILE: No Pension File exists

VETERAN CENSUS: 1888 No / 1890 No / 1894 No

RAY HEREK's BOOK: Pages: 391, 435(Roster)

SPECIAL INFORMATION:
When he enlisted in Northport, Leelanau County, he told the army his home was Cross Village in Emmet County. S-Descriptive Roll, State Archives
He was wounded in the right hand May 18, 1864. S-Herek pg 391
In August 1864 he is listed as "Absent Wounded". S-Descriptive Roll, State Archives
It appears Edward Misisaius was wounded, sent away to a hospital and never heard from again. His fate is a mystery.

LOUIS MISKOGUON 078

Miskoguon, Louis, Charlevoix. Enlisted in company K, First Sharpshooters, Sept. 10, 1863, at Little Traverse, for 3 years, age 30. Mustered Oct. 2, 1863. Missing in action at Petersburg, Va., June 17, 1864. Missing in action July 30, 1864.

Miskoguon, Louis – PRIME NAME, from Adjutant General Record
Miscogeon, Louis – Pension File name
Miscoguom, Louis – 1890 Veteran Census name
Kewagomkun / Kewaguikum, Susan – 1890 Veteran Census (widow)
Kewaquiskum – Diary of Rev. Smith (Dec 18, 1864) S-ccn078a
Kewagoshkom – Letter from Payson Wolf to his wife S-ccn078b
Scogen, L. – Name on Sultana survivors list in a newspaper S-ccn078c

RESIDENCE IN 1863: Charlevoix, Charlevoix County S-AGR
(Note: in 1863, Charlevoix was in Emmet County)
DEATH DATE: July 5, 1870 (*Died near Burgess, Charlevoix County*) S-pf-078
GRAVE: Location Unknown
PENSION FILE: Yes, see Appendix A; 1883 Pension Roll: No
VETERAN CENSUS: 1888 No / 1890 Yes (widow) / 1894 No
RAY HEREK's BOOK: Pages: 369, 399, 403, 435(Roster)

SPECIAL INFORMATION:
The service record of this Civil War soldier is awesome. He survived battles, Andersonville prison, and the Sultana steamboat explosion. Some official army records (Adj. General Record) say he was captured July 30, 1864 at the Battle of the Crater. This is an error, he was captured same time as the other POWs from Company K, on June 17, 1864. S-Herek pg 369
When a list of survivors of the Sultana steamboat disaster was published on the front page of the Detroit Advertiser Tribune on May 24, 1865, his name appeared as "Scogen, L.".
He was mustered out of the army August 9, 1865, at Jackson, Michigan. S-pf-078

His alternate Anishinabe name was **Kewaguikum / Kewaquiskum / Kewagoshkom**.
This name turns up three places: As last name of his widow in 1890 Veteran Census, in the diary of Rev. Smith 12-18-1864, and in a letter Payson Wolf wrote home during the war. S-ccn078b
After Louis died in 1870, his widow married Co. K soldier Simon Sanequaby. S-ccn078d

Louis Miskoguon had strong ties of friendship with a missionary named George N. Smith. When the Reverend Smith started a school for Indians at the Old Wing Mission in Holland, Michigan, Louis was one of the first students (and Payson Wolf was a fellow classmate). When it was time for Louis to marry his sweetheart, Mary "Susan" Ochekwagun, it was the Rev. Smith who conducted the ceremony on Aug 26, 1849. During the Civil War, Smith kept track of Louis and mentions him many times in his personal diary. S-Book: Old Wing Mission, Swierenga, pgs 230; 364

 All Rights Reserved by Chris Czopek

WILLIAM MIXINASAW 079

Mixinasaw, William, Northport. Enlisted in company K, First Sharpshooters, July 9, 1863, at Northport, for 3 years, age 20. Mustered July 20, 1863. Taken prisoner at Petersburg, Va., June 17, 1864. Died in prison at Andersonville, Ga., Oct. 26, 1864. Buried in National Cemetery at Andersonville, Ga. Grave No. 11511.

Mixinasaw, William – PRIME NAME, from Adjutant General Record
Mesenasaw, William – Pension File name
Mackswaser, W. – Name on his Gravestone in Andersonville
Mackswarer, W. – Name on the Atwater List
Mixunasaw, William – Diary of Rev. George N. Smith
Mixernasa, William – Monument in Traverse City

RESIDENCE IN 1863: Northport, Leelanau County S-AGR
DEATH DATE: October 26, 1864 (*Died while a POW at Andersonville Prison*) S-AGR
GRAVE: Georgia / Andersonville National Cemetery, Grave # 11,511
GPS N32.20444 W84.13070

PENSION FILE: Yes, see Appendix A
1883 Pension Roll: No
VETERAN CENSUS: 1888 No / 1890 No / 1894 No
RAY HEREK's BOOK: Pages: 398, 414, 435(Roster)

SPECIAL INFORMATION:
He was captured June 17, 1864 at Petersburg, VA, and sent to Andersonville Prison in Georgia. Two soldiers from Co. K died at Andersonville on the same day: Mixinasaw and Sawbequom.
In Traverse City there is a Veterans Park on 11th Street and Elmwood Avenue. In this park is a block of stone with the names of local men who died in the Civil War. One of the names on this memorial is "William Mixernasa".
William Mixinasaw was a resident of the Northport area, and he belonged to the Rev. George N. Smith's Congregational Mission church. On August 15, 1856, he married Mary Animekewasega (b. 1841, age 15 at the time). The marriage ceremony was conducted by the Rev. Smith.
There were no children who lived to adulthood. When William died, the Rev. Smith filled out the paperwork for Mary to receive a widow's pension. She died a few years later, on March 11, 1868.

NOTES FOR HISTORIANS:
William is mentioned in the diary of Northport missionary Rev. George N. Smith a few times. Smith spells his name **Mixernasa,** Mixunasa, Mixunasaw, and Mixunawaw. S-ccn079a
The Rev. Smith's diary gives the death date for William as August 24 in one place, Oct. 27 in another. Correct death date is October 26, 1864, according to the Atwater List. S-ccn079b

THOMAS MIXONAUBY 080

Mixonauby, Thomas. Enlisted in company K, First Sharpshooters, Sept. 3, 1864, at Grand Rapids, for 1 year, age 39. Mustered Sept. 3, 1864. Died of disease at Detroit, Mich., March 12, 1865.

Mixonauby, Thomas – PRIME NAME, Adjutant General Record
Mixonanby, Thomas – Pension File, aka name
Max-no-bey, Tishpah – 1890 Veteran Census (widow)
Mick se nah bo, Thomas – 1860 US Census
? Ish-ka-buga-ma-ka ? – Herek's book pg. 280-281 (See note below)

RESIDENCE IN 1864: Pentwater, Oceana County S-Descriptive Roll, State Archives
In 1860: Pere Marquette township, Mason County S-1860 US Census
DEATH DATE: March 12, 1865 (*Died of disease in Detroit*) S-AGR
GRAVE: Location Unknown
(*Since he died in Detroit, he is probably buried in a Detroit cemetery*)
PENSION FILE: Yes, see Appendix A
1883 Pension Roll: Yes, Mason County – Custer (widow)
VETERAN CENSUS: 1888 No / 1890 Yes (widow) / 1894 No
RAY HEREK's BOOK: Pages: 418, 435(Roster) Also see pages 280-281

SPECIAL INFORMATION:
He was a 39 year old farmer with a wife named Tishpah who joined the army in Grand Rapids. Soon as he enlisted, Thomas became ill and was constantly in hospitals till his death. S-Shier 246
Official records (Adjutant General Record) say he died in Detroit of disease on March 12, 1865. The hospital he died in was St. Mary's hospital (S-Shier pg 248), but the cemetery he was buried in is a mystery. He should be found in Detroit's Elmwood Cemetery, but his name does not show up in the records. His widow, Tishpah, collected a pension till she died in 1904 at the age of 98.

Herek's book has a story on pages 280-281 of "an old Indian" who joins Co. K in Virginia in the fall of 1864 to take the place of his son, who deserted. **The soldier in Company K most likely to be this "old Indian" is Mixonauby**, but not all the facts fit. Further research needed.

NOTES FOR HISTORIANS:
Thomas Mixonauby was born near Kalamazoo around 1825. He and Tishpah were married in 1856 at the Indian Colony near Leighton in Allegan County. S-Shier pg 248
He enlisted in the army in Grand Rapids, but he was not from Kent County. When he entered the service, he said his home was Pentwater in Oceana County. However, on the 1860 US Census, he and his wife are residents of Pere Marquette township in Mason County.
One of the two pension file index cards for this soldier gives his name "Mixonanby". Obviously the second "N" should be a "U". The other file card (Organization Card) gets the name right.
He may be related to the McSawbe family, but this is not a proven fact. Further research needed.

GEORGE W. MOGAGE 081

Mogage, George W. Enlisted in company K, First Sharpshooters, Feb. 23, 1864, at Kalamazoo, for 3 years, age 23. Mustered Feb. 24, 1864. Joined regiment at Chicago, Ill., March 8, 1864. Absent wounded. No further record.

Mogage, George W. – PRIME NAME, Adjutant General Record
Moogargoe, George W. – Pension File and Gravestone

RESIDENCE IN 1864: Athens township, Calhoun County S-pf-081
Enlisted at Schoolcraft, in Kalamazoo County S-pf-081
In 1866: Athens twp., Calhoun Co. (Reservation of the Nottawaseppi Huron Potawatomi)
From 1866-1873: Jackson, Jackson County S-pf-081
From 1873-1912: Hamilton, Allegan County (Owned an 80 acre farm) S-pf-081

DEATH DATE: May 16, 1923 (*Died at the Grand Rapids Veterans Facility*)
GRAVE: Kent County, Grand Rapids / Grand Rapids Veterans Home Cemetery, Plot 7, Row 19, Grave 8

PENSION FILE: Yes, see Appendix A; 1883 Pension Roll: No

VETERAN CENSUS: 1888 No / 1890 No / 1894 No

RAY HEREK's BOOK: Pages: 391, 435(Roster)

SPECIAL INFORMATION:
His grave in Grand Rapids is on the grounds of the Veterans Home. The gravestone is chipped and cracked, some letters of his name have broken off. The tip of the K in Company K is gone. Few people who glance at this gravestone are aware this man was **the son of a Chief, and he grew up on a reservation in Michigan**. His father was Chief John Mo-gwah-go of the Nottawaseppi Huron band of Potawatomi living on a reservation in Athens township (south of Battle Creek). His tribe had been rounded up and forced to walk the Trail of Death - but his family escaped and quietly returned to Michigan. RESIDENCES: George was born in Constantine, St. Joseph Co., but from childhood till he joined the army, he lived on the reservation in Athens township, Calhoun County. He enlisted in the army at Schoolcraft in Kalamazoo County. **George was wounded in action at Spotsylvania on May 12, 1864.** Sent home on furlough, he had a bad accident and did not recover until the war was over. In 1915 the War Department amended his record to show he was **Discharged for disability June 30, 1864**.
The Grand Rapids Veterans Home was started as an "Old Soldiers Home" for Civil War veterans. George Mogage spent the last years of his life there. **You can find him in the Soldiers Home Record Book as resident number 7132.** He died at the Soldiers Home in 1923 and is buried on the property in Plot 7, Row 19, Grave 8. The main office can give you a map of the cemetery.

DANIEL MWAKEWENAH 082

Mwa-ke-wenah, Daniel, Bear River. Enlisted in company K, First Sharpshooters, Aug. 15, 1863, at Bear River, for 3 years, age 40. Died June 5, 1864, of wounds received in action at Spottsylvania, Va., May 12, 1864. Buried in Brady Hill Cemetery at Saginaw, Mich.

Kewenaw, D. M. A member of company K, First Sharpshooters. Died at Washington, D. C., June 1864, of wounds received in action.

Mwa-ke-wenah, Daniel – PRIME NAME, Adjutant General Record
Mokewenaw, Dan'l – Gravestone name
Kewenaw, D.M. – Adjutant General Record, second entry
Wells, Daniel – aka name for Daniel Mwa-ke-wenah
Mankewenan – See note below

RESIDENCE IN 1863: Bear River (=Petoskey), Emmet County S-AGR
DEATH DATE: June 5, 1864 (*from wounds received in Battle of Spotsylvania, VA*)
GRAVE: Saginaw Co., Saginaw / Brady Hill Cemetery, Section 2D, Lot 12 (Military Stone) *Note: There is a map of this cemetery on the City of Saginaw website*
PENSION FILE: Yes, see Appendix A; 1883 Pension Roll: No
VETERAN CENSUS: 1888 No / 1890 No / 1894 No
RAY HEREK's BOOK: Pages: 36, 148, 155, 156, 390, 435(Roster), 475(burial)

SPECIAL INFORMATION:
Daniel Mwakewenah was Chief of the Bear River band. His home was the Petoskey area.
He signed the Treaty of 1855 as "Mwaw-ke-we-naw".
Member of the Old Mission congregation in Grand Traverse Bay. He was baptized June 4, 1843, at Old Mission by a missionary named Rev. Peter Dougherty and given the name "Daniel Wells". During the Civil War he urged the men of his tribe to volunteer for Company K, and went along with them to share their fate. The Chief was wounded three times at the Battle of Spotsylvania. He died two weeks later in a Washington D.C. army hospital. His death came to the attention of Congressman John Driggs from Michigan, who was aware of the high status Chief Mwakewenah had among Michigan tribes. He arranged for the body to be placed in an expensive casket and sent home to Michigan for burial in Saginaw. For details of his burial, see pages 155, 156, and 475 of Raymond Herek's excellent book. S-ccn082a
When the body of Chief Mwakewenah arrived in Saginaw for burial in 1864, a remarkable story appeared in the local paper. The Chief was called a "hero", and the burial was said to be temporary till arrangements could be made to bring him home. ***See page 219 of this book to read the story.***
The Chief was survived by a widow named Catharine and some small children. Tragically, Catherine died in 1866, leaving her children orphans. The court appointed Ann Rodd, wife of the local interpreter & teacher Daniel Rodd, as the guardian of the two youngest, John and George. The boys received government pension money till they were 16. S-Shier pg 260
There is a Veterans Monument in Hart, Michigan (Oceana County) with his name on it (the name appears as "D.M. Kewenaw").

JOSEPH NARQUAQUOT 083

Nar-qua-quot, Joseph, Pentwater. Enlisted in company K, First Sharpshooters, July 4, 1863, at Elbridge, for 3 years, age 25. Mustered July 11, 1863. Died of disease at Chicago, Ill., Dec. 11, 1863.

Nar-qua-quot, Joseph – PRIME NAME, Adjutant General Record
Nar-quo-quot, Joseph – Pension File name
Naroquaquot, Joseph – Adjutant General Annual Report

RESIDENCE IN 1863: Pentwater, Oceana County S-AGR

DEATH DATE: December 11, 1863 (*Died of disease in Chicago*)

GRAVE: Location Unknown

PENSION FILE: Yes, see Appendix A
1883 Pension Roll: No

VETERAN CENSUS: 1888 No / 1890 No / 1894 No

RAY HEREK's BOOK: Pages: 418, 435(Roster)

SPECIAL INFORMATION:
He was the first member of Company K to die in the war. It happened not in glory on a battlefield, but in a hospital in Chicago. During the Civil War, more soldiers died from disease than bullets. Joseph Narquaquot was one of those victims of illness. **He died from disease while guarding Confederate prisoners at Camp Douglas in Chicago, Illinois.**
Joseph was a young man with a wife named Eliza. Two months after his death, his wife gave birth to a daughter. Mother and daughter moved to Riverton in Mason County, where they applied for pensions from the government. S- Shier pg 262
There is a Veterans Monument in Hart, Michigan (Oceana County) with his name on it.

NOTE TO HISTORIANS:
The location of his grave is a mystery. There were other soldiers in the 1st Michigan Sharpshooters who died while on duty in Chicago, and they were all buried in Rosehill Cemetery, Chicago. But the grave of Joseph Narquaquot does not appear on the cemetery records. Was his body sent home to Oceana County for burial? Further research needed.

 All Rights Reserved by Chris Czopek

JACKSON NARWEGESHEQUABEY 084

Nar-we-ge-she-qua-bey, Jackson, Isabella. Enlisted in company K, First Sharpshooters, June 24, 1863, for 3 years. Mustered July 11, 1863. Missing in action July 30, 1864.

We-ash-kid-ba, Noah. Enlisted in company K, First Sharpshooters, June 24, 1863, at Ionia, for 3 years, age 22. Mustered July 11, 1863. No further record.

Nar-we-ge-she-qua-bey, Jackson – PRIME NAME, Adj. Gen. Record
Nar-ue-ge-she-gwa-by, Jackson – Pension File, aka name
We-ash-kid-ba, Noah – Adj. General Record (same soldier is listed twice)

RESIDENCE IN 1863: Isabella, Isabella County S-AGR

DEATH DATE: Unknown (*Missing in action July 30, 1864, Battle of the Crater*)
GRAVE: Location Unknown

PENSION FILE: Yes, see Appendix A; 1883 Pension Roll: No

VETERAN CENSUS: 1888 No / 1890 No / 1894 No

RAY HEREK's BOOK: Pages: 403, 435(Roster)

SPECIAL INFORMATION:
The last time Jackson was seen alive was at the Battle of the Crater at Petersburg, VA, on July 30, 1864. Official records report him "Missing In Action".
Some records say he was captured on July 30, 1864. S-Descriptive Roll, State Archives
Quita V. Shier, a genealogist in Midland, Michigan, says Jackson Narwegeshequabey was captured on June 17, 1864, and sent to Andersonville Prison. For complete details, see pages 263-264 of her book "Warriors in Mr. Lincoln's Army".
After the war, Jackson's mother received a pension from the government.

NOTE TO HISTORIANS:
This man's name appears twice on the roster of Company K (the Adjutant General Record), first as Jackson Narwegeshequabey, next as Noah Weashkidba. They are the same person.
My theory is this: The Nar- part of Jackson's name was misread by a tired army clerk as "Noa", and mistaken for the first name "Noah".
Compare "Nar_Wegeshequabe_y" to "Noah Weashkidba". Can you see a similarity?
Bad handwriting could explain how his name was misread as a different name. Note also, that both men enlisted on June 24, 1863. Both men came from towns that began with the letter "I" and ended with the letter "a". And both men disappear during the war. Since the army gave a pension to the mother of Jackson Narwegeshequabey, that must be the correct name.

THOMPSON NAUQUAM 085

Nau-quam, Thompson, Charlevoix. Enlisted in company K, First Sharpshooters, June 24, 1863, at Little Traverse, for 3 years, age 23. Mustered July 20, 1863. Deserted at Chicago, Ill.

Nauquam, Thompson – PRIME NAME, Adjutant General Record
Nar-quam, Thompson – Roster in Ray Herek's Book (pg 435)
Naw-o-quom, Thompson – Durant Roll Field Notes #17-pg23

RESIDENCE IN 1863: Charlevoix, Charlevoix County S-AGR
(Note: Back in 1863, Charlevoix was in Emmet County)

DEATH DATE: Unknown

GRAVE: Location Unknown

PENSION FILE: No Pension File exists

VETERAN CENSUS: 1888 No / 1890 No / 1894 No

RAY HEREK's BOOK: Pages: 435(Roster)

SPECIAL INFORMATION:
Thompson's last name was Nau-quam. There was another soldier in Company K with a similar last name. He was Charles Whiteface, and his aka name was Charles Now-quo-um. S-pf-141 (Whiteface pension file)
Compare the two names, **Nau-quam** and **Now-quo-um**. Sound like the same name? There was a one year difference in their ages. Brothers? Cousins? Further investigation needed.

NOTES FOR HISTORIANS:
Durant Roll Field Notes #17-pg23 indicate that in 1907 both he and his wife were dead. There was one surviving child, a son named James Nowquom, Durant Roll No. 4638, Field Notes #26-pg23.

Thompson Nauquam lived in the Charlevoix area. Today Charlevoix (the city) is in Charlevoix County, but at the time of the Civil War, it was in Emmet County. In 1869, Emmet County was split in two, the northern half became Charlevoix County. Keep this in mind when searching for this soldier in census records.

 All Rights Reserved by Chris Czopek

THOMAS NELSON 086

Nelson, Thomas, Taymouth. Enlisted in company K, First Sharpshooters, as Corporal, May 18, 1863, at Isabella, for 3 years, age 27. Mustered May 26, 1863. Transferred to Veteran Reserve Corps Sept. 16, 1864.

Nelson, Thomas – PRIME NAME, all records agree on the spelling

RESIDENCE IN 1863: Taymouth, Saginaw County S-AGR
DEATH DATE: August 4, 1877 S- Saginaw County Death Records, Liber 2A-H231, Record 56
GRAVE: Saginaw Co., Taymouth twp. / Taymouth Indian Cemetery (Military Stone)
(This cemetery is also called Wheaton Indian cemetery, look for it near Birch Run)
PENSION FILE: Yes, see Appendix A; 1883 Pension Roll: No
VETERAN CENSUS: 1888 No / No / 1894 No
RAY HEREK's BOOK: Pages: 204, 205, 412, 435(Roster)

SPECIAL INFORMATION:
Thomas Nelson was one of the first men to enlist in Company K.
He was the son of a Chief and a natural born leader, so they immediately made him a Corporal. Wounded in action at the Battle of Spotsylvania. His right leg was hit by a bullet, and it never healed right. He was lame for the rest of his life. S-Herek pg 204
Transferred to the VRC (Veteran Reserve Corps). He served in the VRC 14th Regt., Co. A, till the end of the war. S-pf-086

A description of the life of Thomas Nelson can be found in the book by Quita V. Shier titled "Warriors in Mr. Lincoln's Army", pages 268-271. It is the best source for detailed information about his family. According to this book, Thomas was the son of the well known and respected Chippewa Chief David Shoppenagonce. This is the same "Chief Shoppenagon" who moved to Grayling, Michigan, and is still remembered there; the town's hotel is named after him.
The Chief's son, Thomas Nelson, had a close connection with the **Taymouth Indian Mission** (located near Birch Run in Saginaw County). He was married there before the war, lived there after his return from the army, and is buried in the Taymouth Indian Cemetery.
He married Mary Fisher (aka Wawwawsawmoquay) in 1859. They had five children.
Life was not easy for this disabled veteran. His battle-wound never healed properly and he had difficulty walking. He tried working as a laborer and lumberman, but found it hard to hold a job. His wife Mary died in 1875, leaving him to care for the children. Thomas Nelson knew his own life was coming to an end, so he asked his brother-in-law, Peter Gruet Jr. (who also had enlisted in Company K) to look after his children when he was gone. After Thomas Nelson died in 1877, Peter Gruet Jr. took the youngest children into his home and raised them as his own. S-Shier pg 270

NOTE TO HISTORIANS:
The VRC was a branch of the army. Its ranks were filled up with soldiers who had minor injuries that prevented them from marching and fighting with their original regiments.
By transferring to the VRC, these men could still be serving in the army, but their jobs were easier, such as hospital guards. The VRC freed up able bodied men for service on the battlefield.

JOSEPH NESOGOT 087

Ne-so-got, Joseph. Enlisted in company K, First Sharpshooters, June 5, 1863, at Isabella, for 3 years, age 27. Mustered June 16, 1863. No further record.

Williams, Joseph, Isabella. Enlisted in company K, First Sharpshooters, June 5, 1865, at Isabella, for 3 years, age 27. Mustered June 16, 1863. Taken prisoner at Petersburg, Va., June 17, 1864. Discharged at Detroit, Mich., June 20, 1865.

Ne-so-got, Joseph – PRIME NAME, Adjutant General Record
Shoyat, Jos – Gravestone name, correct reading
Shovet / Shovat, Jos – Common misreading of the gravestone name
Williams, Joseph - Adjutant General Record, second entry

RESIDENCE IN 1863: Isabella township, Isabella County S-AGR
After the war, he lived in Taymouth twp., near Birch Run in Saginaw Co. S-ccn087a
DEATH DATE: July 10, 1873 S-Govt. Gravestone Contract Record
GRAVE: Saginaw Co., Taymouth twp. / Taymouth Indian Cemetery (Military Stone)
Name on gravestone is Jos Shoyat (Note: This cemetery is also called Wheaton Indian Cemetery, located on Seymour Road near intersection with Birch Run Road in Saginaw County)
PENSION FILE: No Pension File exists
VETERAN CENSUS: 1888 No / 1890 No / 1894 No
RAY HEREK's BOOK: Pages: Nesogot 435(Roster); Williams 288, 399, 435(Roster), 495

SPECIAL INFORMATION:
Joseph Nesogot aka Williams was captured and sent to the infamous Andersonville Prison. He survived – but his health was broken. He died several years after the end of the war.
Joseph Nesogot grew up on a farm on the Isabella reservation. S-1860 US Census
After the war he moved to Taymouth township in Saginaw Co. He died there in 1873. S-ccn087a
I am certain Joseph Nesogot is the same person as Joseph Williams, for the following reasons: Joseph Nesogot enlisted at Isabella on June 5, 1863. So did Joseph Williams. Both men were 27 years old and disappear after the war. Also, Nesogot's father was William Nesogot. S-ccn087b
One of the mysteries of Company K is a gravestone in the Taymouth Indian cemetery.
It is a military stone for a Company K soldier named JOS SHOYAT. Name is not on the roster of Company K, so who is Joseph Shoyat? I say this is the grave of Joseph Nesogot. His brother, Samuel Ne-So-Got, is buried close by. An expert on the families of Taymouth township has told me that Joseph Nesogot was sometimes called Sogot / Shogot. In my opinion, bad handwriting on the order form for the gravestone turned the word SHOGOT into SHOYAT.
He is on the Gruett Roll, page 165, as Joseph Ne-se-got. (Father = William Ne-se-got).
ERROR IN OFFICIAL RECORDS: Notice the Adjutant General Record says Joseph Williams enlisted June 5, 1865. This is obviously an error, and the correct year of his enlistment is 1863.

WILLIAM NEVEAUX 088

Nevarre, William. Enlisted in company K, First Sharpshooters, as Corporal, July 8, 1863, at LaPoint, for 3 years, age 18. Mustered July 20, 1863. Mustered out at Delaney House, D. C., July 28, 1865.

Neveaux, William – PRIME NAME, from his pension file
Nevarre, William – Adjutant General Record name

RESIDENCE IN 1863: La Pointe, Ashland County, Wisconsin S-AGR
(*Note: La Pointe is a town on Madeline Island*)
In 1890: Bayfield, Ashland County, Wisconsin S-1890 Veteran Census
(*Note: The ferry boat to Madeline Island leaves from Bayfield, Wisconsin*)
In 1914: La Pointe, Wisconsin S-pf-088

DEATH DATE: October 16, 1914 (*Died at La Pointe, Wisconsin*) S-pf-088
GRAVE: Wisconsin, Ashland County, Madeline Island, La Pointe /Buried in the Old Catholic Mission Cemetery (Military stone) S-ccn088a
PENSION FILE: Yes, see Appendix A
1883 Pension Roll: No
VETERAN CENSUS: 1888 No / 1890 Yes / 1894 No

RAY HEREK's BOOK: Pages: 435(Roster)

SPECIAL INFORMATION:
William Neveaux enlisted in Company K at La Pointe, Wisconsin. S-AGR
He served on the Color Guard for a while. S-DR
In 1864 he was seriously ill. After recovering, he served as a hospital attendant. S-ccn088b
When the war ended, he returned to Madeline Island in Wisconsin, married, and raised a family.
Member of the A. E. Burnside G.A.R. Post No. 249, Bayfield, Wisconsin. S-Shier pg 274
After his death in 1914, his second wife, Elizabeth, received a widows' pension for 25 years.
At the time of her death on July 30, 1939, she was the LAST SURVIVING WIDOW OF A COMPANY K SOLDIER. S-pf-088

NOTES FOR HISTORIANS:
He enlisted as "Nevarre", but the name "Neveaux" is on his pension file, and on his gravestone.
Since William Neveaux enlisted at La Pointe, Wisconsin, and many years later he died at La Pointe (as evidenced by his grave there), we can assume La Pointe was his home.
The town of La Pointe is on MADELINE ISLAND in Ashland County, Wisconsin.
This Island has a special place in the traditions and teachings of the Anishinabek.
See pages 102-103 of the book by Edward Benton-Banai "The Mishomis Book – The Voice of the Ojibway" for the fascinating story of Madeline Island.

WILLIAM NEWTON 089

Newton, William, Pentwater. Enlisted in company K, First Sharpshooters, July 4, 1863, at Pentwater, for 3 years, age 17. Mustered July 11, 1863. Mustered out at Delaney House, D. C., July 28, 1865.

Newton, William – PRIME NAME, Adjutant General Record name
Knowlton, William C. – Obituary S-Traverse City Record Eagle, January 28, 1922

RESIDENCE IN 1863: Pentwater, Oceana County S-AGR
In 1886: Burnham, Manistee County S-pf-011 (pension file for Amos Ashkebugnekay)
In 1889: Northport, Leelanau County S-G.A.R. Post No. 399 records, State Archives
In 1890: Leland township, Leelanau County S-1890 Veteran Census
In 1894: Traverse City, Grand Traverse County S-1894 Veteran Census
In 1922: Yuba, Grand Traverse County S-see below

DEATH DATE: January 23, 1922
(*Died in Williamsburg, Grand Traverse County* S-pf-089)

GRAVE: Grand Traverse County, near Acme / Yuba Cemetery
Yuba cemetery is on US-31 near Acme. Easiest way to find Yuba cemetery is to look on the west side of the highway for the road leading to Sayler Park. Yuba cemetery is nearby, on the opposite side of the highway, the east side. His gravestone is a bronze plaque that lies flat on the ground.

PENSION FILE: Yes, see Appendix A
1883 Pension Roll: No

VETERAN CENSUS: 1888 No / 1890 Yes / 1894 Yes

RAY HEREK's BOOK: Pages: 435(Roster), 495

SPECIAL INFORMATION:
One of the 15 soldiers from Company K captured at Petersburg, VA on June 17, 1864.
He was sent to Andersonville Prison – and survived!
His health was remarkable. The only one of the seven Andersonville survivors who returned to the regiment in time to be mustered out at Delaney House on July 28 with the rest of Company K (All the others were discharged from army hospitals). On the 1890 Veteran Census, he reported no disabilities. And he outlived almost every other member of Company K.
Newton was member # 34 of the Woolsey G.A.R. Post No. 399, Northport, Leelanau County
He died in Williamsburg, Grand Traverse County (S-pf-089), but was a resident of Yuba.
His obituary gave his name as "William C. Knowlton". S-Traverse City Record Eagle, January 28, 1922
His wife, Angeline Newton, was one of the last three suriviving widows of a Company K soldier. She died Feb. 4, 1936. (Susan Pequongay d. July 20, 1939. Elizabeth Neveaux d. July 30, 1939).

LEMUEL R. NICHOLS 090

Nichols, Lemuel R., Fayette. Enlisted in company C, First Sharpshooters, Oct. 23, 1863, at Woodland, for 3 years, age 21. Mustered Oct. 24, 1863. Promoted Sergeant. Promoted First Lieutenant, company K, Nov. 20, 1864. Mustered Dec. 14, 1864. Mustered out at Delaney House, D. C., July 28, 1865.

Note: Before enlisting in the 1st Michigan Sharpshooters, he served in the Michigan 7th Infantry:

Nichols, Lemuel R., Oakland County. Enlisted in company H, Seventh Infantry, Aug. 7, 1861, at Oxford, for 3 years, age 19. Mustered Aug. 22, 1861. Discharged for disability at Washington, D. C., June 30, 1862. Died Jan. 29, 1893. Buried at Owosso, Mich.

Nichols, Lemuel R. – PRIME NAME, from Pension File
Nichols, Robert – 1886 booklet for Reunion of 1st Mich. SS S-ccn090a

RESIDENCE:
In 1861: Oakland County, near Oxford S-AGR for 7th Michigan Infantry
In 1863: Fayette, Hillsdale County S-AGR
In 1890: Owosso, Shiawassee County S-1890 Veteran Census

DEATH DATE: January 29, 1893 S-AGR for 7th Michigan Infantry
GRAVE: Shiawassee County, Owosso, Oak Hill Cemetery, Section L
PENSION FILE: Yes, see Appendix A; 1883 Pension Roll: No
VETERAN CENSUS: 1888 No / 1890 Yes / 1894 No
RAY HEREK's BOOK: Pages: 281, 425(Roster)

SPECIAL INFORMATION:
White officer (1st Lieutenant) in Company K.
No relation to Lt. Colonel Asahel Nichols. S-ccn090b
Was promoted from Sergeant in Co. C to 1st Lieutenant of Co. K.
He was 1st Lieutenant of Company K from Nov. 20, 1864 to the end of the war.
He attended the 1886 Reunion of the 1st Michigan Sharpshooters.

NOTE TO HISTORIANS:
He first served for a short time in the 7th Michigan Infantry, Co. H.
There is a booklet at State of Michigan Library in Lansing (call number E 514.5, 1st, F57 .Z) printed for the 1886 Reunion of the 1st Michigan Sharpshooters. On page 15 is a list of the members of Company K at the reunion. Lt. Nichols appears twice: First as "Lieut. Robert Nichols", and a second time as "Nichols, Lemuel R."
He grew up in Fayette township, Hillsdale County. His wife was named Sarah M. Nichols.
Further research needed to see if any wartime letters have been passed down in his family, letters that would give details about the men in Company K.

KIRKE W. NOYES 091

Noyes, Kirk W., South Haven. Enlisted in company D, First Sharpshooters, as First Sergeant, Dec. 27, 1862, at South Haven, for 3 years, age 32. Mustered March 31, 1863. Wounded in action at Spottsylvania, Va., May 12, 1864 Commissioned First Lieutenant, company K, July 15, 1864. Mustered July 15, 1864. Wounded in action at Petersburg, Va., July 30, 1864. Wounded in action and taken prisoner at Peebles Farm, Va., Sept. 30, 1864. Paroled Feb. 22, 1865. Commissioned Captain company B, April 11, 1865. Mustered June 14, 1865. Mustered out at Delaney House, D. C., July 28, 1865. Present residence, South Haven, Mich.

Noyes, Kirke W. – PRIME NAME, from Pension File
Noyes, Kirk W. – Adjutant General Record

RESIDENCE IN 1862: South Haven, Van Buren County S-AGR
In 1890: Paw Paw, Van Buren County S-1890 Veteran Census
In 1907: South Haven, Van Buren County S-AGR
DEATH DATE: August 8, 1920 S-pf-091
GRAVE: Van Buren Co., South Haven / Lakeview Cemetery (Military stone)
Buried in Block 2, Lot 23, Grave #4
Easiest way to find his grave is to first find where the cemetery exits onto Bailey street. Look there for a large, old Beech tree. His grave lies in the shade of that tree.
PENSION FILE: Yes, see Appendix A; 1883 Pension Roll: No
VETERAN CENSUS: 1888 No / 1890 Yes / 1894 Yes
RAY HEREK's BOOK: Pages: 258, 259, 310-311, 323, 350, 390, 402, 405, 426(Roster)

SPECIAL INFORMATION:
White officer (1st Lieutenant) in Company K for about ten weeks.
He enlisted as First Sergeant of Company D, and that is the unit on his gravestone. He was later promoted to First Lieutenant and transferred to Company K. He served as an officer in Co. K for ten weeks, from July 15, 1864 until he was wounded and taken prisoner at Peebles Farm, Va., on Sept 30, 1864. After he was released from prison, he was promoted to Captain of Company B. After the war he lived for a while in Paw Paw, working as Postmaster, Justice of the Peace, and several other jobs in Van Buren County government. He was commander of the local GAR Post.

NOTE TO HISTORIANS:
Seems the correct spelling of his first name is Kirke with "e" at end, but you will find it commonly spelled Kirk with no "e". S-ccn091a
Further research needed to see if any wartime letters have been passed down in his family, letters that would give details about the men in Company K. This former officer lived till 1920 – did anyone write down his memories about leading Company K in the Battle of the Crater?

JAMES OHBOWAKEMO 092

Oshburn, James R. Enlisted in company K, First Sharpshooters, Feb. 1, 1864, at Kalamazoo, for 3 years, age 26. Mustered Feb. 2, 1864. Mustered out at Delaney House, D. C., July 28, 1865.

Ohbowakemo, James – PRIME NAME, from Pension File
Oshburn, James R. – Adjutant General Record
Aubwakoum, James – 1890 Veteran Census
Olwayquonske, James – Name on roster in Ray Herek's Book, pg 435

RESIDENCES:
Before the war, visited the Chippewa village Shimnecon on the Grand River S-pf-092
In 1864: Enlisted in Schoolcraft, Kalamazoo County S-Descriptive Roll, State Archives
In 1887: Sarnia, Ontario, Canada (Living on an Indian reserve there) S-pf-092
In 1890: Oregon twp. & Elba twp., Lapeer Co. S-1890 Veteran Census & S-pf-124 (Tabasash)
In 1891-1895: Nottawa twp., Isabella County S-pf-124 (pension file for Francis Tabasash)

DEATH DATE: December 21, 1895 (*Died in Nottawa township, Isabella County)* S-pf-092
GRAVE: Location Unknown (see note below)
PENSION FILE: Yes, see Appendix A; 1883 Pension Roll: No
VETERAN CENSUS: 1888 No / 1890 Yes / 1894 No
RAY HEREK's BOOK: Pages: 435(Roster)

SPECIAL INFORMATION:
His pension file hints that he was a Chippewa member of one of the Grand River Bands in Ionia County near Shimnecon, but no definite information is given. Further research needed.
In 1887 he was living in Canada on a reservation near Sarnia. Then he moved back to Michigan and lived in Lapeer County. The last years of his life were spent in Isabella County. S-pf-092
He was never well after the war. Seems he contracted consumption (tuberculosis) while in the army. For the rest of his life he was too weak to do a full day's work. Married four times. S-pf-092
When his third wife died, James was left with a baby daughter he was unable to care for. His friend from army days, John Waubenoo of Weidman, took care of the child in his home. S-Shier pg 283
The little girl, named Mary, grew up and married Amos Elk on December 4, 1899. S-pf-092

NOTES FOR HISTORIANS:
The name "Oshburn" appears only on the Adjutant General Record, and the Descriptive Roll.
The name **Jacob P.Yuncker** appears in the pension file as guardian for James' child. Jacob was the Nottawa township supervisor. One of his duties was to distribute money to families in need. He testified that when James died in 1895, the township paid for the coffin and the funeral. S-pf-092
Since this soldier was buried by the county, there should be a record of expenses. It may be possible to locate his grave by searching the Nottawa township records between the dates December 21, 1895 and January 21, 1896. Those records may say where he was buried.

LEON OTASHQUABONO 093

O-tash-qua-bon-o, Leon, Little Traverse. Enlisted in company K, First Sharpshooters, Aug. 22, 1863, at Northport, for 3 years, age 23. Mustered Oct. 21, 1863. Promoted Corporal July 3, 1865. Mustered out at Delaney House, D. C., July 28, 1865. Died April 3, 1902, at Harbor Springs, Mich.

Otashquabono, Leon – PRIME NAME, from Pension File
O-tash-qua-bon-o, Leon – Adjutant General Record
Otashquabona, Leon – State Death Record name
Otasquabona, Leon – 1894 Veteran Census
Otosgabench, Leone – 1890 Veteran Census, aka name
Odesheweiad, Leo – Name in newspaper obituary S-Petoskey Record 4-9-1902

RESIDENCE IN 1863: Little Traverse (=Harbor Springs), Emmet County S-AGR
In 1842: Canada (Born in Canada) S-State Death Record
In 1889: Ironton, Charlevoix County S-pf-078 (pension file for Louis Miskoguon)
In 1890: Eveline township, Charlevoix County S-1890 Veteran Census

DEATH DATE: April 3, 1902 S-State Death Record

GRAVE: Emmet County, Harbor Springs / Lakeview Cemetery (Military stone)
(*He is buried in the Catholic section, which is called Holy Childhood Cemetery*)
GPS Lat/Long: 45.44058, -84.99838

PENSION FILE: Yes, see Appendix A; 1883 Pension Roll: No

VETERAN CENSUS: 1888 No / 1890 Yes / 1894 Yes
RAY HEREK's BOOK: Pages: (Preface xii), 367, 435(Roster)

SPECIAL INFORMATION:
Leon was born in Canada January 6, 1842 S-State Death Record. (His metal grave marker says born 1841).
When he enlisted in Northport, he told the army his home was Little Traverse (= Harbor Springs).
Leon was member # 111 of the Baxter G.A.R. Post No. 119, Charlevoix, Charlevoix County.
He also was a member of the Bowen G.A.R. Post No. 340, Ironton, Charlevoix County.
There is a photograph of Leon wearing his G.A.R. badge and holding his G.A.R. hat. Look for it in the DeLand family files at the Burton Historical Collection, Detroit.
Some sources say Leon is buried in Holy Childhood Cemetery. This is the name of the Catholic section of Lakeview Cemetery, Harbor Springs.
His grave has two markers: The standard Civil War military gravestone, and an unusual metal memorial featuring an angel gazing downward at his gravestone.

MARCUS OTTO 094

Otto, Marcus, Isabella. Enlisted in company K, First Sharpshooters, May 18, 1863, at Isabella, for 3 years, age 19. Mustered May 26, 1863. Discharged for disability at Alexandria, Va., Aug. 18, 1864.

Otto, Marcus – PRIME NAME, from Adjutant General Record
Kenewegeshik, Marcus – aka name, from his Pension File
Waw-bay-gaw-me-sking, Marcus – signature on Land Patent (see below)

RESIDENCE IN 1863: Isabella, Isabella County S-AGR
In 1900 - 1904: Weidman, Nottawa township, Isabella County S-pf-094

DEATH DATE: March 26, 1904 S-pf-094
GRAVE: Either buried on his farm near Weidman, Isabella County S-ccn094a
or **Nottawa Indian Cemetery (aka Redman Cemetery)**, Nottawa twp., Isabella Co.
S-State Death Record (see note below)

PENSION FILE: Yes, see Appendix A
1883 Pension Roll: Yes, Isabella Co. – Mt. Pleasant (Loss of right arm above elbow)
VETERAN CENSUS: 1888 Yes / 1890 Yes / 1894 Yes
RAY HEREK's BOOK: Pages: 346, 409, 435(Roster)

SPECIAL INFORMATION:
Marcus lost his right arm in the Battle of Spotsylvania, VA. S-pf-094 (DeLand letter dated Mar 4 '65)
There is a tintype photograph of Marcus taken after he was discharged from the army.
You can see this picture in the book "AUBE NA BING", by Simon Otto, published in 1988.
Simon is the grandson of Marcus Otto.
Marcus Otto was member #59 of Wa-Bu-No G.A.R. Post No. 250, Mt. Pleasant, Isabella Co. He joined this Post in 1884, the first year it was formed, but he was not a charter member.
Marcus and his brother Solomon lived on the Isabella reservation near Mt. Pleasant, Michigan. When a recruiter for Company K came to the reservation, the two brothers enlisted in the army together. Each received a $25 bounty (S-pf-094). His older brother, Solomon, died in the war. Marcus married his brother's widow. S-Gruett Roll pg 365 And he inherited his brothers land grant from the government in 1871, signing for it with the name "Marcus Waw-bay-gaw-me-sking". His father was named Waw be gaw me skum. For details of his family tree, see Gruett Roll pg 365. The tragic story of the death of Marcus Otto was the front page story of the Mt. Pleasant newspaper "Central Times" on April 1, 1904.
There is a controversy over the location of his grave: According to some descendants, Marcus Otto was buried on his farm in Nottawa township, near Weidman in Isabella County. There is no gravestone to mark the spot, but his grandchildren can point to the exact place. S-ccn094a
However, a State Death Record on file in Lansing says Marcus was buried in Nottawa Cemetery, in Nottawa township (today known as **Nottawa Indian Cemetery**, aka Redman Cemetery, located on Denver Road near Vandecar Road). No gravestone has been found for him in this cemetery.

SOLOMON OTTO 095

Otto, Solomon, Isabella. Enlisted in company K, First Sharpshooters, May 18, 1863, at Isabella, for 3 years. Mustered May 26, 1863. Absent, wounded Sept., 1864. No further record.

Kar-ga-yar-sang, Solomon. Enlisted in company K, First Sharpshooters, May 18, 1863, at Isabella, for 3 years, age 24. Mustered May 26, 1863. No further record.

Otto, Solomon – PRIME NAME, Adjutant General Record name
Kargayarsang, Solomon – Adjutant General Record, second entry
Kaw-gay-aw-sung – Name on the Gruett Roll, page 365

RESIDENCE IN 1863: Isabella, Isabella County S-AGR
DEATH DATE: May 1865 (*Died shortly after May 2nd discharge from the army*)
GRAVE: Location Unknown
PENSION FILE: No Pension File exists
RAY HEREK's BOOK: Pages: 404, 435(Roster)

SPECIAL INFORMATION: Solomon and his brother, Marcus, lived on the Isabella reservation. The two brothers enlisted in the army together. Each received a $25 bounty. S-ccn095a
Solomon Otto was wounded on Sept. 30, 1864, at the Battle of Peebles Farm, VA. S-Herek pg 404
The following story was passed down from father to son in the Otto family: Solomon was very sick at a hospital in Detroit. He was dying, so they put him on a train for home. He never arrived. It is supposed he died somewhere along the way. S-ccn095b
Army records confirm this family story. The papers in his service file at National Archives show that Solomon was discharged as a disabled soldier on May 2nd, 1865, at Harper Hospital in Detroit. He left for home and was never seen again. It appears that somewhere along the way to the reservation in Isabella County, he became ill and died.
Solomon was married. After the war, Marcus married Solomon's widow. S-Gruett Roll pg 365

NOTES FOR HISTORIANS: This soldier appears twice on the Adjutant General Record, first as Solomon Kargayarsang, second as Solomon Otto. Here is the evidence:
1) Official records say nothing about Kargayarsang other than his enlistment and age, which match the enlistment information for Solomon Otto. 2) A roster of Company K found in State Archives dated July 22, 1863, shows the name Marcus Otto, but does not show his brother Solomon Otto. It does, however, show the name Solomon Kargayarsang. Another roster in State Archives dated December 14, 1863, shows both Marcus Otto and Solomon Otto, but not Solomon Kargayarsang.
3) The Gruett Roll, page 365, shows that Marcus had an older brother named "Kaw-gay-aw-sung". This name is too close to "Kargayarsang" to be coincidence. They are definitely the same person.
THE 1871 LAND GRANT - According to terms of the 1855 & 1864 Treaties with the Chippewa, a parcel of land was reserved for Solomon. When the government finally honored the agreement in 1871, Solomon was dead. His brother Marcus was the legal heir, and he signed the land patent as "Marcus Waw-bay-gaw-me-sking". He is identified by mistake as "son of grantee". S-ccn095c

JOHN PAKEMABOGA 096

Pa-ke-mabo-ga, John. Enlisted Dec., 1863, at Chicago, Ill. Supposed to have been killed July 30, 1864.

Pa-ke-mabo-ga, John – PRIME NAME, Adjutant General Record

RESIDENCE IN 1863: Unknown

DEATH DATE: July 30, 1864 (*KIA at the Battle of the Crater, Petersburg VA*) S-AGR
(Note: KIA = Killed In Action)

GRAVE: Location Unknown

PENSION FILE: No Pension File exists

VETERAN CENSUS: 1888 No / 1890 No / 1894 No

RAY HEREK's BOOK: Pages: 402, 435(Roster)

SPECIAL INFORMATION:
Ray Herek's book, page 402, says he died from wounds received in Battle of The Crater.
The army records have very little to say about this man.

NOTE TO HISTORIANS:
Some researchers think that John Pakemaboga is the same person as John Penegwah, a man who does not appear on the roster of Company K and is known only from one pension file (# 281,385). I respectfully disagree. The mysterious John Penegwah is indeed on the roster, but he appears under another name, John Kinewahwanipi. The proof is simple – compare the names of the parents:
John Penegwah's parents = Joseph Banagua and Margaret Negonbequay.
John Kinewahwanipi's parents = Joseph Penegwa and Naganbequa.

The mystery surrounding John Pakemaboga remains. I suspect he may be **John Wabesis**, who was reported killed in the same battle. But that is just a hunch. Further research needed.

DANIEL PEMASSEGAY 097

Pemas-se-gay, Daniel, Isabella. Enlisted in company K, First Sharpshooters, May 18, 1863, at Isabella, for 3 years, age 24. Mustered May 26, 1863. Deserted at Dearborn, Mich., July 5, 1863.

Pemas-se-gay, Daniel – PRIME NAME, Adjutant General Record

RESIDENCE IN 1863: Isabella, Isabella County S-AGR

DEATH DATE: Unknown

GRAVE: Location Unknown

PENSION FILE: No Pension File exists

VETERAN CENSUS: 1888 No / 1890 No / 1894 No

RAY HEREK's BOOK: Pages: 435(Roster)

NOTE FOR HISTORIANS:
The Gruett Roll, page 232B, shows a **Daniel Tay-baw-se-gay** living in Canada with a wife and child. Is this the same person as Daniel Pemassegay? Further research needed.

JACKO PENAISWANQUOT 098

Pe-nais-now-o-quot, Jacko, Northport. Enlisted in company K, First Sharpshooters, June 1, 1863, at Northport, for 3 years, age 40. Mustered June 22, 1863. Taken prisoner at Petersburg, Va., June 14, 1864. Died in prison at Andersonville, Ga.

Penaiswanquot, Jake "Jacko" – PRIME NAME, from Pension File
Pe-nais-now-o-quot, Jacko – Adjutant General Record name
Pe-nay-se-waw-naq-quot – Durant Roll Field Notes #22-pg41
Penaeswonoquot, Jocko – Adjutant General Annual Report, aka name
Pas-an-o-quoit, Jaco – Pension File, aka name
Recalt, S. – Name on gravestone at Andersonville National Cemetery
Ricott, S. – Name on the Atwater List S-ccn098a
Shako / Shake – His name in the Rev. Smith Diary (see below)

RESIDENCE IN 1863: Northport, Leelanau County S-AGR
DEATH DATE: November 1, 1864 S-ccn098a
Died while a POW at Andersonville Prison, and is buried under the name S. Ricott S-ccn098a
GRAVE: Georgia / Andersonville National Cemetery, Grave #11,705
Gravestone says "S. Recalt" (There is a gravestone with his name in Virginia, see note below)
PENSION FILE: Yes, see Appendix A; 1883 Pension Roll: No
VETERAN CENSUS: 1888 No / 1890 No / 1894 No
RAY HEREK's BOOK: Pages: 288, 393, 399, 414, 435(Roster)

SPECIAL INFORMATION:
He joined the army at age 40, leaving behind a wife and 8 children. They never saw him again. The missionary Rev. George N. Smith calls him "Shako" in his diary. (Some have misread this word as "Shake".) His wife was called "Mrs. Jacko", and his children adopted "Jacko" or "Shako" as their last name. People from Peshawbestown named Shako/Jacko are descendants of this soldier. Two of his sons joined Company K: John and Louis (aka John Jacko and Nattahmenoting Jacko).
Error in records: Jacko Penaiswanquot was captured on June 17, not on June 14. S-pf-098
He was one of 15 soldiers from Company K captured and sent to the infamous Andersonville Prison. Official Army records say he died there, but the location of his grave has been a mystery. Lansing historian Roger Winthrop has determined that S. Ricott / Recalt who died on Nov. 1, 1864 is Jacko Penaiswonquot, and I agree. Jacko's pension file says he died October 30, 1864, but that date is based on the memory of his fellow soldiers. The Atwater death date is more accurate.
There is a gravestone in Poplar Grove National Cemetery, near Petersburg, VA, with the name Jaco Penaswonquot. This is an error, the name should have been Pe-to-zo-on-oquette.
Herek's book says Jacko Penaiswanquot was an Indian doctor. This is an error. The testimony of his widow in her pension file explains how the error was made.
He appears in Durant Roll Field Notes #22-pg41 under the name Pe-nay-se-waw-naw-quot.

 All Rights Reserved by Chris Czopek

MARK PESHEKEE 099

Pe-she-kee, Mark, Isabella. Enlisted in company K, First Sharpshooters, May 19, 1863, at East Saginaw, for 3 years, age 30. Mustered May 26, 1863. Absent wounded Aug., 1864. No further record.

Pe-she-kee, Mark – PRIME NAME, from Adjutant General Record

RESIDENCE IN 1863: Isabella, Isabella County S-Descriptive Roll, State Archives

DEATH DATE: Unknown

GRAVE: Location Unknown

PENSION FILE: No Pension File exists

VETERAN CENSUS: 1888 No / 1890 No / 1894 No

RAY HEREK's BOOK: Pages: 148, 391, 435(Roster)

SPECIAL INFORMATION:
His name means "Buffalo".
Mark Peshekee joined the army in East Saginaw, but Isabella received credit for draft purposes, meaning Isabella was probably his home in 1863. S-Descriptive Roll, State Archives
At the Battle of Spotsylvania, he was shot in the left shoulder and sent to a hospital. S-Herek pg 148
Afterwards he disappears from army records. There is no doubt he died in the war.

NOTES FOR HISTORIANS:
Back in the 1850's, the leader of the La Pointe Band in Wisconsin was "Chief Buffalo". He is famous in Wisconsin history, and his life is well documented. His Anishinabe name is Beshekee, and it means "Buffalo". Compare the chief's name Beshekee to the soldiers' name Peshekee – they are nearly identical. It is tempting to suppose a family connection between these two, based on the similarity of their names. However, there is absolutely no proof of a connection between the Company K soldier from Isabella and famous "Chief Buffalo" in Wisconsin.

 All Rights Reserved by Chris Czopek

ALBERT PESHERBAY 100

Pisherbay, Albert, Charlevoix. Enlisted in company K, First Sharpshooters, June 18, 1863, at Little Traverse, for 3 years, age 25. Mustered July 20, 1863. Discharged at Detroit, Mich., Aug. 11, 1865.

Pesherbay, Albert – PRIME NAME, from Pension File
Pisherbay, Albert – Adjutant General Record name
Pasharba, Albert – 1894 Veteran Census
Pershabay, Albert – Newspaper obituary S-Petoskey Evening News, June 6, 1908
Peshawbey, Albert – Cemetery records have this name

<u>RESIDENCE IN 1863:</u> Charlevoix, Charlevoix County S-AGR
In 1890: Bear Lake township, Charlevoix County S-1890 Veteran Census
In 1894: Bear Creek township, Emmet County S-1894 Veteran Census

<u>DEATH DATE:</u> August 29, 1903 (*Died in Bear Creek twp., Emmet Co.*)S-State Death Record
<u>GRAVE:</u> Emmet County, Petoskey / Greenwood Cemetery (Unmarked grave)
(*Definitely buried in the Potters Field section, but exact spot is unknown.
In July 2009, the cemetery dedicated a monument to all the veterans buried in Greenwood who do not have a gravestone. Albert's name is on that monument.*)

<u>PENSION FILE:</u> Yes, see Appendix A; 1883 Pension Roll: No

<u>VETERAN CENSUS:</u> 1888 No / 1890 Yes / 1894 Yes

<u>RAY HEREK's BOOK:</u> Pages: 435(Roster)

SPECIAL INFORMATION:
His father's name was "Black Wolf Peshabe". There is a strong similarity between the name <u>Peshabe</u> and the town in Leelanau County named <u>Peshawbestown</u>. However, I can find no definite evidence of a connection. S-ccn100a
At Petersburg, the same exploding shell that killed Petros Awanakwad caused injuries to Albert Pesherbay that plagued him in the final years of his life. S-pf-100
There is no obituary for his death in 1903, but he is mentioned in his widow's obituary as "**Albert Pershabay, who was an honored veteran**". S-Petoskey Evening News, June 6, 1908, pg 3
He is buried in an unmarked grave in Greenwood Cemetery, Petoskey. There is a monument in Greenwood to honor the veterans there without marker stones. Albert's name is on that monument.

NOTE TO HISTORIANS:
When you search for him in census records, remember Charlevoix was in Emmet Co. until 1869. Also, there was a Bear Lake township in Charlevoix Co., and a Bear Creek township in Emmet Co. Albert lived in both places. (Bear Lake township existed then, but does not exist today).

CHARLES PORCELLAY 101

Porsley, Charles, Burt. Enlisted in company K, First Sharpshooters, July 6, 1863, at Little Traverse, for 3 years, age 25. Mustered July 20, 1863. Wounded in action Sept., 1864. Discharged Dec. 17, 1864.

Porcellay, Charles – PRIME NAME, from Pension File
Porsley, Charles – Adjutant General Record name
Parceley, Charles – Descriptive Roll, State Archives

RESIDENCE IN 1863: Burt, Cheboygan County S-AGR, and Adj. Gen. Annual Report

DEATH DATE: Unknown

GRAVE: Location Unknown

PENSION FILE: Yes, see Appendix A; 1883 Pension Roll: No

VETERAN CENSUS: 1888 No / 1890 No / 1894 No

RAY HEREK's BOOK: Pages: 409, 435(Roster)

SPECIAL INFORMATION:
Member of the Burt Lake band in Cheboygan County.
Enlisted in the army at Little Traverse (= Harbor Springs).
Charles Porcellay was in all the big battles of 1864: The Wilderness, Spotsylvania, the savage fight outside of Petersburg on June 17th, and the infamous battle at the Crater on July 30th.
His official service record says he was wounded in action. S-AGR
Went to the hospital and was diagnosed with Consumption (Tuberculosis). Mustered out at L'Ouverture Hospital in Alexandria on Dec 26, 1864. Went back to Michigan on a train full of happy soldiers going home on Christmas leave. There is no doubt he arrived home at Burt Lake. We know this for a fact, because he applied for a pension in August 1865. The application was mailed from Burt Lake. What happened after that is a mystery. He vanishes from all records.

NOTES FOR HISTORIANS:
Charles was one of three men from the Burt Lake who joined the army during the Civil War. Their names are: Simon Sanequaby, Charles Porcellay, and the Chief's son Simon Kejikowe. All three men served together in Company K.
There was a soldier named Charles Porceley in Company I, who was taken prisoner and died in Salisbury Prison. Do not confuse him with Charles Porcellay in Company K. These are two different people.

JACOB PRESTAWIN 102

Prestawin, Jacob. Enlisted in company K, First Sharpshooters, Nov. 18, 1864, at Traverse City, for 1 year, age 24. Mustered Nov. 18, 1864. Absent wounded Sept., 1864. No further record.

Prestawin, Jacob – PRIME NAME, all sources agree with spelling

RESIDENCE IN 1864: Bingham, Leelanau County S-Descriptive Roll, State Archives

DEATH DATE: Unknown (*Died while in the army during the Civil War.*)

GRAVE: Location Unknown

PENSION FILE: No Pension File exists

VETERAN CENSUS: 1888 No / 1890 No / 1894 No

RAY HEREK's BOOK: Pages: 404, 435(Roster)

SPECIAL INFORMATION:
This soldier is something of a mystery. Not much is said about him in the official army records. Last report says he was wounded at Peebles Farm on September 30, 1864. S-Herek pg 404
After that he disappears. **The fate of this man is one of the unsolved mysteries of Company K.**

NOTE TO HISTORIANS:
ERROR IN THE RECORDS: The Adjutant General Record says Jacob Prestawin enlisted in November 1864, and was absent wounded in September 1864. See the error? Either the enlistment date is wrong, or the casualty report day is wrong. Further research needed to straighten this out.

Could this man be the mysterious Nattahmenoting Jacko?
In the pension file for Nattahmenoting Jacko (# 158,777) is a note that says *"The name Nai-tah-me-no-ting Jocko is not borne on rolls of Co. K ... but the name **Pestawin Jacko** appears on rolls ... enlistment and description corresponds with that on the Discharge Certificate of the first-named man ..."* S-pf-052 (pension file of Nattahmenoting Jacko #158,777)
This could solve an old mystery. There is no doubt that Nattahmenoting (aka Louis Penaiswonquot, son of Jacko Penaiswonquot and brother of John Jacko) enlisted in Company K. So why is his name missing on the official army roster of Company K? Perhaps it is on the roster after all, appearing as Jacob Prestawin. Compare the enlistment date, place, and physical descriptions of the two men:
Nattahmenoting Jacko – Enlisted 1864 Nov 18 / Home: Leelanau area / Age: 24 / Height: 5ft 9-1/2
Prestawin Jacko – Enlisted 1864 Nov 18 / Home: Leelanau area / Age: 24 (when discharged) / Height: 5ft 9-1/2
See the similarity? Further research needed to confirm if these two men are the same person.

JAMES QUOBOWAY 103

Quar-bo-way, James, Isabella. Enlisted in company K, First Sharpshooters, May 18, 1863, at Isabella, for 3 years, age 32. Mustered May 26, 1863. Discharged Dec. 28, 1864.

Quoboway, James – PRIME NAME, from Pension File
Quar-bo-way, James – Adjutant General Record name
Oquarboway, James – Adjutant General Annual Report
Quabaway, James – 1894 Veteran Census

RESIDENCE IN 1863: Isabella, Isabella County S-AGR
In 1883: Mt. Pleasant, Isabella County S-1883 Pension Roll
In 1894: Deerfield township, Isabella County S-1894 Veteran Census

DEATH DATE: January 4, 1906 (*Died in Deerfield twp., Isabella Co.*) S-State Death Record

GRAVE: Location Unknown (see note below)

PENSION FILE: Yes, see Appendix A
1883 Pension Roll: Yes, Isabella County – Mt. Pleasant (Wounded left hip)

VETERAN CENSUS: 1888 No / 1890 No / 1894 Yes

RAY HEREK's BOOK: Pages: 435(Roster)

SPECIAL INFORMATION:
James Quoboway was wounded in action at the battle of Spotsylvania. A bullet seriously damaged his left hip, and the wound would not heal. He was discharged from the army as a disabled veteran. For the next forty years he collected a disability pension. S-ccn103a

NOTE TO HISTORIANS:
The location of his grave is a mystery, but we have these clues:
The State Death Record says he was buried in Isabella township on January 6, 1906. The same record says he was buried by an undertaker named H.A. Foster of Mt. Pleasant. Find the records of this undertaker, and you will find the place where James is buried.

JOHN RUBBERDEE 104

Rubberdee, John. Enlisted in company K, First Sharpshooters, Sept. 26, 1864, at Rockland, for 3 years, age 25. Mustered Oct. 20, 1864. Joined regiment at Peebles' Farm, Va., Nov. 4, 1864. Mustered out at Delaney House, D. C., July 28, 1865.

Rubberdee, John – PRIME NAME, Adjutant General Record
Rabberden, John – Descriptive Roll, State Archives
Rabberdee, John – Descriptive Roll, aka name

RESIDENCE IN 1864: Rockland, Ontonagon County S-Descriptive Roll, State Archives
(*Rockland is in Michigan's Upper Peninsula*)

DEATH DATE: Unknown

GRAVE: Location Unknown

PENSION FILE: No Pension File exists

VETERAN CENSUS: 1888 No / 1890 No / 1894 No

RAY HEREK's BOOK: Pages: 435(Roster)

SPECIAL INFORMATION:
When the war was over, John Rubberdee paid six dollars so he could keep his army rifle and take it home. S-ccn104a
He was from the Upper Peninsula (Ontonagon County). S-Descriptive Roll, State Archives
Enlisted the same day, same place as Peter Amderling.
Army records indicate he went through the war with no serious illness or injury. After his discharge from the army he went home to the Upper Peninsula and disappeared.

NOTE TO HISTORIANS: The last name is probably spelled wrong. It is not an Anishinabe name. I have been told by an expert his **correct name is probably Robideu, or Robideaux**, names of prominent French-Canadian families that settled in Wisconsin and Michigan's Upper Peninsula. How could Robideu become Rubberdee? I suspect bad handwriting is to blame.

AARON SAHGAHNAHQUATO 105

Sahgahnahquato, Aaron. Enlisted in company K, First Sharpshooters, Feb. 14, 1865, at Grand Rapids, for 1 year, age 21. Mustered Feb. 14, 1865. Discharged at Detroit, Mich., Aug. 11, 1865. Present residence, Leland, Mich.

Sahgahnahquato, Aaron – PRIME NAME, Adjutant General Record
Sargonquatto, Aaron – Pension File name & gravestone name
Sargaswongeto, Aaron – G.A.R. Post No. 399 records
Pequongay, Aaron – Pension File, aka name
Pequongway, Aaron – Label on photograph in museum S-ccn105a

RESIDENCE IN 1865: Leelanau twp., Leelanau Co. S-Descriptive Roll, State Archives
In 1861: Manistee "Worked in a sawmill in Manistee before the war" S-pf-105
In 1888: Gills Pier [Onumunese Village], Leelanau County S-1888 Veteran Census
In 1890: Leelanau township, Leelanau County S-1890 Veteran Census
In 1897: Northport, Leelanau County S-G.A.R. Post No. 399 records at State Archives
In 1907: Leland, Leelanau County S-AGR "Present Residence, Leland, Mich."

DEATH DATE: June 4, 1916 (*Died in Northport from heart disease*) S-State Death Record
GRAVE: Leelanau County, Omena /Omena Hillcrest Cemetery (Military Stone)
Hillcrest Cemetery is on Kalchik Road. Directions: Take M-22 to Omena, turn left (West) on County Hwy 626 (=Tatch Rd) and follow it West for a short distance till you see Kalchik road. Turn right (North) on Kalchik Road and follow it for half a mile. Cemetery will be on the right side.
PENSION FILE: Yes, see Appendix A; 1883 Pension Roll: No
VETERAN CENSUS: 1888 Yes / 1890 Yes / 1894 Yes
RAY HEREK's BOOK: Pages: 290, 435(Roster)

SPECIAL INFORMATION:
One of three men recruited by Payson Wolf when he was home after release from Andersonville. His gravestone in Omena is unique – It has two names! It is a standard Military stone with the name Aaron Sargonquatto at top, and at the bottom are the words "Known as Aaron Pequongay". His pension file says he was discharged from the Army July 28, 1865, with the rest of Co. K. **The Grand Rapids Public Museum has a photograph of Aaron Pequongway and his family. Look for it in the museum's Native American heritage display. Picture taken around 1915.** His wife, Susan, was the sister of Sgt. Charles Allen, a Company K soldier killed in battle. Susan died July 20, 1939 – one of the last surviving widows of a Company K soldier. Aaron's uncle was Company K soldier Joseph Kakakee. (S-pf-061, pension file for Joseph Kakakee). Aaron belonged to the Rev. Dougherty's Presbyterian church in Omena, and is mentioned many times in church records. **Aaron was member #37 of the Woolsey G.A.R. Post No. 399, Northport, Michigan.** There is a newspaper story written in 1955 that says Aaron was with General Sherman on the famous March to the Sea. This is not true. Company K only fought in Virginia. S-ccn105b

SIMON SANEQUABY 106

Sanequaby, Simon, Burt. Enlisted in company K, First Sharpshooters, as Corporal, June 8, 1863, at Little Traverse, for 3 years, age 25. Mustered June 22, 1863. Promoted Sergeant. Mustered out at Delaney House, D. C., July 28, 1865.

Sanequaby, Simon – PRIME NAME, from Adjutant General Record
Singoby, Simon – Pension File name
Sanegoby, Simon – G.A.R. records, Post No. 106
Stgobe, Simeon – Name on State Death Record
Say-nin-gwaw-day – Name on Durant Roll S-Durant Roll Field Notes #5-pg31

RESIDENCE IN 1863: Burt Lake, Cheboygan County S-Descriptive Roll, State Archives

DEATH DATE: June 27, 1903 S-State Death Record
(*Died from Typhoid Fever in Hayes township, Charlevoix County*) S-State Death Record

GRAVE: Emmet County, Resort twp. / The old Indian cemetery near Bay Shore
(*Buried in an unmarked grave in the "Indian Cemetery" located on Old US-31 Hwy, between Townline Rd and Horton Bay Road. This cemetery is near the Kinglet Nature Preserve. Some old records named it Carpenter Cemetery after the nearest post office. See note below)*

PENSION FILE: Yes, see Appendix A; 1883 Pension Roll: No
VETERAN CENSUS: 1888 No / 1890 No / 1894 No
RAY HEREK's BOOK: Pages: 281, 435(Roster)

SPECIAL INFORMATION:
One of three men from the Burt Lake band who enlisted into Company K during the Civil War. Their names are: Simon Sanequaby, Charles Porcellay, and the Chief's son Simon Kejikowe. Simon Sanequaby enlisted in Company K as a Corporal, was later promoted to Sergeant.
Simon was member #33 of the Washington G.A.R. Post No. 106, Cross Village, Emmet Co.
He was a member of the Burt Lake band and lived on Colonial point. He moved to Hayes township in Charlevoix County one or two years before the infamous Burt Lake burn-out in 1900.
His wife, Mary Singoby, was the daughter of Angelus Madwess. S-State Death Record
The Durant Roll says Simon and Mary had no surviving children. S-Durant Roll Field Notes #5-pg31
Simon died of Typhoid Fever in the summer of 1903. S-State Death Record.
His comrade from Company K, William Isaacs, gave the information for his State Death Record.

NOTE TO HISTORIANS:
The State Death Record says his place of burial is "Carpenter". Carpenter was a crossroads village with a Post Office in Emmet County located at the intersection of Horton Bay Road and Martinchek Road. It no longer exists. The closest town to where old Carpenter stood is Bay Shore.

SASHKOBANQUOT 107

Sash-ko-ban-quot, Pinconning. Enlisted in company K, First Sharpshooters, May 18, 1863, at Isabella, for 3 years, age 45. Mustered May 26, 1863. Accidentally killed at Camp Douglas, Chicago, Ill., Dec. 27, 1863.

Sash-ko-ban-quot – PRIME NAME, from Adjutant General Record
Sash-ke-bouquot – History of Saginaw County, pg 368 S-ccn107a
Shosh-bono-quot – Name in letter by Agent George W. Lee S-ccn107b
Aish-ke-baw-quot – Name on tribal roll S-ccn107c
Kobanquot, Shash – Pension File name

RESIDENCE IN 1863: Pinconning, Bay County S-AGR
DEATH DATE: December 27, 1863
GRAVE: Location Unknown (see note below)
PENSION FILE: Yes, see Appendix A; 1883 Pension Roll: No
VETERAN CENSUS: 1888 No / 1890 No / 1894 No
RAY HEREK's BOOK: Pages: 77, 420, 435(Roster)

SPECIAL INFORMATION:
He was one of the oldest soldiers in Company K.
While the 1st Michigan Sharpshooters were guarding Confederate prisoners in Chicago, he went to town on a pass and was accidentally killed, crushed between two railroad cars. S-Herek pg 77
The location of his grave is a mystery. All the other soldiers in the 1st Michigan Sharpshooters who died in Chicago are buried in Rosehill Cemetery. But the cemetery burial records do not show the name of this soldier. Further research needed.
His name appears on a monument in Bay City. On Memorial Day, 2011, his name was added to the Veterans Monument in front of the Bay County building. To visit this monument, park in the parking lot of Wirt Library, located at 500 Center Street in Bay City. The county building is across the street.

NOTE TO HISTORIANS:
After his death, two women, Mary and Saline, claimed to be his widow and applied for a pension. Mary applied June 20, 1864, application # 54,722. Saline applied July 13, 1865, application # 101,734. Saline was previously recognized as widow and received the soldier's back pay. But in the end, the government recognized Mary as the true widow

Years later, Etoqua, "the widow of Shosh-bono-quot", was sick and in dire straits. The U.S. Indian Agent for Michigan tried to help her obtain a widow's pension. For details, see the 1879 – 1880 correspondence between Indian Agent George W. Lee and the Commissioner of Indian Affairs in Washington D.C.

ADAM SAWBEQUOM 108

Saw-be-come, Adam, Pentwater. Enlisted in company K, First Sharpshooters, July 4, 1863, at Pentwater, for 3 years, age 28. Mustered July 11, 1863. Taken prisoner at Petersburg, Va., June 17, 1864. Died in prison at Andersonville, Ga., Oct. 26, 1864. Buried in National Cemetery at Andersonville, Ga. Grave No. 11508.

Sockum, A. Member of company K, First Sharpshooters. Died Oct. 26, 1864. Buried at Andersonville, Ga.

Saw-be-quom, Adam – PRIME NAME, from Pension File
Saw-be-come, Adam – Adjutant General Record
Sockum, A. – Adj. General Record (same soldier is listed twice)
Sockem, A. – Atwater List of soldiers who died at Andersonville
Saw-be-come, Xedan – Gravestone name (see note below)
Sarrbequon, Na-ta-ma-may – 1883 Pension Roll (Mother)
Sa-Qua-On – Rev. George N. Smith diary, 12-18-1864
Osah-o-quah-yon – Letter home from Payson Wolf S-ccn108a

RESIDENCE IN 1863: Pentwater, Oceana County S-AGR
DEATH DATE: October 26, 1864 (*Died while a POW at Andersonville Prison*) S-AGR
GRAVE: Georgia / Andersonville National Cemetery, Grave # 11,508
GPS N32.20444 W84.13070
PENSION FILE: Yes, see Appendix A
1883 Pension Roll: Yes, Mason Co. – Marble (his mother received a pension)
VETERAN CENSUS: 1888 No / 1890 No / 1894 No
RAY HEREK's BOOK: Pages: 399, 414, 435(Roster) /A. Sockum: 414, 435(Roster)

SPECIAL INFORMATION:
ERROR IN RECORDS: Same soldier is listed twice on Adjutant General Record's roster.
"Adam Saw-be-come", and "A. Sockum" are the same man (compare death date and place).
He was captured June 17, 1864 at Petersburg, VA, and sent to Andersonville Prison in Georgia.
Correct death date at Andersonville is October 26, 1864. S-Atwater List
His gravestone says "Xedan" instead of Adam. This is a clerical error, not an Anishinabe name.
The Atwater list spells his name "Sockem, A.". Adjutant General Record spells it "Sockum, A."
DEATH DATES: Atwater list says Oct. 26 , but the Adj. Gen. Annual Report says Oct. 21,
and the Descriptive Roll in State Archives says Nov. 26. Trust the date on the Atwater list.
Two soldiers from Co. K died at Andersonville on the same day, Mixinasaw and Sawbequom.
There is a Veterans Monument in Hart, Michigan (Oceana County) with his name on it.

ANTOINE SCOTT 109

Scott, Antoine, Pentwater. Enlisted in company K, First Sharpshooters, as Corporal, July 4, 1863, at Pentwater, for 3 years, age 22. Mustered July 11, 1863. Mustered out at Delaney House, D. C., July 28, 1865.

Scott, Antoine – PRIME NAME, from Adjutant General Record
Le Croix, Antoine – Pension File, aka name
Wi yaw be mind, Antoine – Pension File, aka name

RESIDENCE IN 1863: Pentwater, Oceana County S-AGR
DEATH DATE: December 10, 1878 S-GENDIS Death Index
GRAVE: Location Unknown (*Died in Custer township, Mason Co.* S-GENDIS Death Index)
PENSION FILE: Yes, see Appendix A; 1883 Pension Roll: No
VETERAN CENSUS: 1888 No / 1890 No / 1894 No
RAY HEREK's BOOK: Pages: 227, 251, 321, 342, 368, 435(Roster), 500

SPECIAL INFORMATION:
Born at Mackinac around 1841. His father was named Mitchell Le Croix. S-pf-109
He enlisted in the army under the name "Antoine Scott" on the 4th of July. S-AGR
Antoine Scott was a genuine Civil War hero. Repeatedly cited for bravery, he was promoted to Sergeant. He really distinguished himself during the tragic Battle of the Crater July 30, 1864. Official Army Records say: "Instead of screening himself behind the captured works, this soldier stood boldly up and deliberately fired his piece until the enemy was close upon him, when, instead of surrendering, he ran the gauntlet of shot and shell and escaped". This action delayed the enemy advance, buying time for the Sharpshooters to retreat to the safety of the Union line. When Major General John G. Parke, commander of the 9th Army Corps, heard this story, he recommended Antoine for the **Medal of Honor** (S-ccn109a). For technical reasons, the medal was never awarded. See Raymond Herek's book, pages 251, 342, 368 & 500 for details.
He was twice recommended for MOH. The second time was for heroism on April 2, 1865, during an attack against the enemy line at Petersburg, VA. See Ray Herek's book, page 321, for details.

When the war ended, Antoine Scott returned home to Michigan. At the time of his death in 1878, he was living in Mason County, working as a laborer, and married. S-GENDIS Death Index
He died at the age of 37 from illness (inflammation of the lungs). There is no record of where he is buried, but since he lived near Custer, Michigan, it is almost certain he rests in the section of Custer Riverside Cemetery known as "Indian Burial Ground".
In 1890 his mother, Petabun, asked the government for a pension. The papers she sent with her pension-application are the best source of information on the family of Antoine Scott.

NOTE TO HISTORIANS: According to genealogist Quita V. Shier, there was a blood connection between Antoine Scott and another soldier in Company K named Michael Wayaubemind. The two men were half-brothers ! Same mother (her name was Petabun) different fathers. For complete details, see Quita's book "Warriors In Mr. Lincoln's Army", pg 332 and pg 435.

JOSEPH SEYMOUR 110

Seymour, Joseph. Enlisted in company K, First Sharpshooters, Sept. 24, 1864, at Greenland, for 3 years, age 20. Mustered Oct. 20, 1864. Joined regiment at Peebles' Farm, Va., Nov. 4, 1864. Mustered out at Delaney House, D. C., July 28, 1865.

Seymour, Joseph – PRIME NAME, all spellings agree

RESIDENCE IN 1864: Greenland, Ontonagon County S-Descriptive Roll, State Archives (*Greenland is in Michigan's Upper Peninsula*)

DEATH DATE: Unknown

GRAVE: Location Unknown

PENSION FILE: No Pension File exists

VETERAN CENSUS: 1888 No / 1890 No / 1894 No

RAY HEREK's BOOK: Pages: 435(Roster)

SPECIAL INFORMATION:
Very little is known about this soldier.
He was recruited in Michigan's Upper Peninsula (Ontonagon County) late in the war as a replacement for the heavy losses suffered by Company K during the summer of 1864.
We know he served on the front lines at Petersburg and other places for ten months.
At the end of the war he was honorably discharged with the rest of the regiment.
Then he vanishes into history.

CHARLES SHABENA 111

Shabema, Charles, LaCroix. Enlisted in company K, First Sharpshooters, as Wagoner, June 11, 1863, at LaCroix, for 3 years, age 23. Mustered July 20, 1863. Mustered out at Delaney House, D. C., July 28, 1865.

Shabena, Charles – PRIME NAME, from Pension File
Shabema, Charles – Adjutant General Record name

RESIDENCE IN 1863: LaCroix (=Cross Village), Emmet County S-AGR

DEATH DATE: September 13, 1866 S-pf-111 (Pension file of Charles Shabena)

GRAVE: Location Unknown

PENSION FILE: Yes, see Appendix A
1883 Pension Roll: No

VETERAN CENSUS: 1888 No / 1890 No / 1894 No

RAY HEREK's BOOK: Pages: 435(Roster)

SPECIAL INFORMATION:
Charles Shabena was married in Cross Village on March 4, 1862 to Mary Ann Kijebinessi. This was a church wedding performed by the resident priest, Father Zorn. No children. One year later, he joined the army (June 11, 1863).
Charles Shabena was not your ordinary infantryman. He was a wagoner.
A "wagoner" is a wagon driver. Company K was given a wagon for hauling supplies on a march. Charles Shabena was recruited to be the driver of that wagon. He survived the war, but came home in the Summer of 1865 with a bad case of bronchitis that never went away. One year later, he died of pneumonia. His funeral service was conducted by the resident priest, Father Zorn.

NOTE TO HISTORIANS:
The Adjutant General Record spells the name "Shabema" with an M. His pension file spells it "Shabena" with an "N". In my opinion, his pension file is correct.
NOTE: *There was a famous Odawa war chief named Shabbona. His name is sometimes spelled Shabena. Because of the similarity of names, there is a slight possibility that Charles Shabena of Company K was related to this Chief. But this is not a proven fact. Further research needed.*

WINFIELD S. SHANAHAN 112

Shanahan, Winfield S., Cassopolis. Enlisted in company E, First Sharpshooters, as Corporal, March 6, 1863, at Niles, for 3 years, age 18. Mustered March 31, 1863. Missing in action at Cold Harbor, Va., June 12, 1864. Promoted Second Lieutenant, company K, March 7, 1865. Mustered April 10, 1865. Mustered out at Delaney House, D. C., July 28, 1865.

Shanahan, Winfield S. – PRIME NAME, all sources agree on spelling

RESIDENCE IN 1863: Cassopolis, Cass County S-AGR
(*The 1850 US Census says his home was in La Grange township, Cass County*)

DEATH DATE: 1904 (Widow applied for pension in November, 1904)

GRAVE: Location Unknown

PENSION FILE: Yes, see Appendix A; 1883 Pension Roll: No

VETERAN CENSUS: 1888 No / 1890 No / 1894 No

RAY HEREK's BOOK: Pages: 334, 393, 428(Roster), 495

SPECIAL INFORMATION:
White officer (2nd Lieutenant) in Company K.
Winfield S. Shanahan was the 2nd Lieutenant of Company K during the last months of the war, from March 7, 1865 to July 28, 1865.
He started out as a Corporal in Company E. Captured at Cold Harbor on June 12, 1864.
Paroled and returned to regiment December 2, 1864. S-Herek pg 393
Three months after his release from prison, he was promoted to 2nd Lieutenant of Company K. Very little is known about his life, other than he grew up on a farm in La Grange twp., Cass Co., and his wife was named Carrie Shanahan. We know Carrie was living in Massachusetts in 1904 when she applied for a widow's pension. Further research needed to see if any letters have been passed down in his family, letters that would give details about the men in Company K.

CHARLES SHAW 113

Shaw, Charles, Pentwater. Enlisted in company K, First Sharpshooters, July 4, 1863, at Elbridge, for 3 years, age 19. Mustered July 11, 1863. Mustered out at Delaney House, D. C., July 28, 1865.

Shaw, Charles – PRIME NAME, Adjutant General Record
Boo Shaw, Charles – Find A Grave website

RESIDENCE IN 1863: Pentwater, Oceana County S-AGR

DEATH DATE: December 23, 1899 S-Find A Grave website

GRAVE: Oceana County, Hart / Hart City Cemetery (Military stone)
Look for his gravestone near the stack of cannon balls

PENSION FILE: No Pension File exists

VETERAN CENSUS: 1888 No / 1890 No / 1894 No

RAY HEREK's BOOK: Pages: 281, 341, 435(Roster)

SPECIAL INFORMATION:
He volunteered to join Company K on the 4th of July.
Promoted to Corporal. Went through the war without wounds or serious illness.
Charles Shaw was the only member of Company K to serve on the Color Guard.
S-Herek pg 341
(Note: The Color Bearer was the soldier chosen to carry the Regiment flag in battle. Since he could not carry a weapon, a "Color Guard" was appointed to march with him and protect the flag. It was a great honor to be chosen for the Color Guard.)
THE GRAVE OF CHARLES SHAW is easy to find. Simply drive around the City Cemetery of Hart, Michigan, till you see a stack of cannon balls, then look at the gravestones closest to the stack.

NOTE TO HISTORIANS:
A long time ago, a Civil War Cannon was placed in the cemetery as a memorial to the Civil War veterans buried there. During WWII the cannon was donated to a scrap drive, but the stack of cannon balls that stood beside it still remains. The closest grave to this stack is the Charles Shaw grave.

JOSEPH SHAWANESE 114

Shaw-au-ase, Joseph, Chesaning. Enlisted in company K, First Sharpshooters, June 13, 1863, at East Saginaw, for 3 years, age 23. Mustered June 16, 1863. Discharged June 27, 1865.

Shaw-an-ese, Joseph – PRIME NAME, from Pension File
Shaw-au-ase, Joseph – Adjutant General Record name
Shawnes, Jos. – Name on Gravestone
Shaw-an-ax, Joseph – History of Saginaw County, Chapman, pg 368
Shon-am-niss, Joseph – 1890 Veteran Census, aka name
Shaw, Joseph – Mich. Dept. of Health Death Index (GENDIS)

RESIDENCE IN 1863: Chesaning, Saginaw County S-AGR
From 1875 - 1878: Indian Town, near St. Charles, Saginaw County S-pf-114
DEATH DATE: May 20, 1878 S-pf-114
(*GENDIS, the online Mich. Dept. of Health Death Index, says Joseph Shaw died 6/16/1878 of consumption. This may be the same person as Joseph Shaw-an-ese. Further research needed*)
GRAVE: Saginaw County, St. Charles / Riverside Cemetery (Military stone)
Name on gravestone is "Jos. Shawnes"
Note: Highway M-52 divides the cemetery in half; the grave is located in the east half.
PENSION FILE: Yes, see Appendix A; 1883 Pension Roll: No
VETERAN CENSUS: 1888 No / 1890 Yes (widow) / 1894 No
RAY HEREK's BOOK: Pages: 391, 435(Roster)

SPECIAL INFORMATION:
He was the tallest man in Company K ! When he enlisted, he measured 6ft 4-1/2 in. S-pf-114
Before the war, Joseph lived in the Chesaning and St. Charles area of Saginaw County. S-pf-114
He earned his living by hunting and trapping. S-pf-114
Wounded in the hand at the Battle of the Wilderness. He was sent to a hospital in Alexandria, VA, for treatment. One finger on his left hand amputated. While in the hospital he was diagnosed with tuberculosis. Discharged & sent home to Michigan. Died from lung disease in 1878. S-pf-114
ERROR IN RECORDS: There are four different dates given for his discharge. In his pension file is a letter from the War Department saying he was discharged August 18, 1864 for disability. That is probably the correct date. He was not discharged in May or June of 1865.
The Michigan Dept. of Health has a death record for "Joseph Shaw", an Indian who died in St. Charles of consumption on June 16, 1878. I strongly suspect this is Joseph Shawanese.
S-GENDIS, Ledger page 53, Record No. 270

NOTE TO HISTORIANS:
He lived in Indian Town near St. Charles. Located south of St. Charles on the Shiawassee river, then East of the bridge where Furgus road crosses the Shiawassee river. East of this bridge is where "Indian Town" was. Another way of describing the location is: "Between the river and the corner of Furgus and Turner roads". S-Two local historians in Chesaning, MI

 All Rights Reserved by Chris Czopek

JOSEPH SHAWONOSANG 115

Shaw-au-os-ang, Joseph, Brown Island. Enlisted in company K, First Sharpshooters, June 18, 1863, at Little Traverse, for 3 years, age 18. Mustered July 20, 1863. Discharged for disability at Detroit, Mich., Jan. 17, 1865.

Shawonosang, Joseph – PRIME NAME, from Pension File
Shaw-au-os-ang, Joseph – Adjutant General Record name
Shaw-em-as-ang, Joseph – "Father Abraham's Children" by Woodford
Shanernuskin, Joseph – Pension File, aka name

RESIDENCE IN 1846: Born in Canada on Manitoulin Island c. 1846 S-pf-115
In 1863: Beaver Island, Charlevoix County (See note below)
In 1867: Moved to Attica township, Wyoming County, New York S-pf-115
In 1879: Pinconning, Bay County S-pf-115

DEATH DATE: Unknown
(His pension file at National Archives is missing some pages, so there is no death date)
GRAVE: Location Unknown
PENSION FILE: Yes, see Appendix A; 1883 Pension Roll: No
VETERAN CENSUS: 1888 No / 1890 No / 1894 No
RAY HEREK's BOOK: Pages: 391, 409, 435(Roster)

SPECIAL INFORMATION:
Born in Canada on Manitoulin Island. Moved to Beaver Island and worked for James F. Cable. When Cable retired in 1867 and moved to Attica, New York, Joseph went with him. S-pf-115
Joseph later returned to Michigan. In 1879 he was living near Pinconning, Bay County. S-pf-115
Joseph was severely wounded in the left leg at the Battle of Spotsylvania May 12, 1864.
The book "Father Abraham's Children" by Frank B. Woodford, on pages 212-213, says Joseph went to Harper Hospital after the war for "surgical treatment" and stayed there six months.
His pension file at National Archives has pages missing; no date & place of death is given.
DISCHARGE DATE: Adjutant General Record says "discharged for disability" on Jan 17, 1865. However, his discharge paper says December 17, 1864 at Harper Hospital, Detroit. Since he applied for a pension December 27, 1864, I say trust the date on his discharge paper.
NOTE TO HISTORIANS:
Both the Adjutant General Record & Descriptive Roll say his residence in 1863 was Brown Island in Manitou County. This is a problem, because there is no Brown Island in Manitou County. (Note: Manitou County existed from 1855 to 1895 and included Beaver Island plus a few small islands. None of those small islands are named Brown Island.)
I have come to the conclusion that "Brown" is a clerical error, and should read "Beaver".
This soldier came from Beaver Island in Manitou County. His pension file confirms this, saying that he was working for James F. Cable on Beaver Island before he joined the army.
On Beaver Island there is a Veterans Memorial; Joseph's name will be put on one of the bricks.

JOHN SHEGOGE 116

She-go-ge, John, Pentwater. Enlisted in company K, First Sharpshooters, July 4, 1863, at Elbridge, for 3 years, age 21. Mustered July 11, 1863. Wounded in action June 17, 1864. Discharged June 5, 1865.

Shegoge, John – PRIME NAME, Adjutant General Record
Shegog, John – Pension File, aka name
Sheyoy [Shegog], John – Pension File name, see note below

RESIDENCE IN 1863: Pentwater, Oceana County S-AGR
In 1865: Pentwater, Michigan S-pf-116

DEATH DATE: May 26, 1869 S-Shier pg 348

GRAVE: Location Unknown (*Died in Pentwater, Oceana County* S-Shier pg 348)

PENSION FILE: Yes, see Appendix A; 1883 Pension Roll: No

VETERAN CENSUS: 1888 No / 1890 No / 1894 No

RAY HEREK's BOOK: Pages: 396, 435(Roster)

SPECIAL INFORMATION:
His name "Shegoge" means "skunk".
John Shegoge joined the army on the 4th of July.
Wounded in action near Petersburg, VA on June 17, 1864.
Right arm amputated below the elbow.
He took a long time to recover – Records say he was in one army hospital after another from June 1864 to June 1865 – twelve months! S-ccn116a
Discharged after the war ended, at Harper Hospital, Detroit.

NOTE TO HISTORIANS:
The correct spelling of his name is controversial. His pension file consistently spells the name "Sheyoy", but this appears to be a clerical error. I believe his application for a pension was misread by a pension department clerk, and the letter "g" was misread as letter "y". Then the mistake was copied over and over. **Correct spelling of his last name is Shegoge.**

JOHN B. SHOMIN 117

Shomin, John B. S., LaCroix. Enlisted in company K, First Sharpshooters, Aug. 17, 1863, at Grand Traverse, for 3 years, age 20. Mustered Oct. 21, 1863. Taken prisoner at Petersburg, Va., June 17, 1864. Returned to regiment May 6, 1865. Discharged at Washington, D. C., July 3, 1865. Present residence, Mackinac Island, Mich.

Shomin, John B. – PRIME NAME, from Pension File
Shomin, John B.S. – Adjutant General Record name
Shoman, John B. – 1894 Veteran Census
Shomin, John Baptiste – Baptismal Name S-ccn117a

RESIDENCE IN 1863: LaCroix (=Cross Village), Emmet County S-AGR
In 1890: Mackinac Island, Mackinac County S-1890 Veteran Census
DEATH DATE: September 16, 1913 S-pf-117
GRAVE: Mackinac Co., Mackinac Island / St. Anne Cemetery (Military Stone)
(There is a gravestone for John B. Shomin in Lakeview Cem., Harbor Springs, see note below)
PENSION FILE: Yes, see Appendix A; 1883 Pension Roll: No
VETERAN CENSUS: 1888 Yes / 1890 Yes / 1894 Yes
RAY HEREK's BOOK: Pages: Preface xii, 288, 367, 399, 435(Roster), 495

SPECIAL INFORMATION:
Company K had three men named SHOMIN. John B. Shomin and Louis Shomin were brothers. John O. Shomin was their cousin. All three grew up together in Cross Village.
John B. Shomin was captured and sent to Andersonville Prison. He survived.
He was member #38 of the Fenton G.A.R. Post No. 258, St. Ignace, Mackinac County.

IMPORTANT NOTE: Company K had two men named "John Shomin"; they were cousins. An army clerk named one "John", and the other "John B".
After the war, John lived in Harbor Springs, and John B. moved to Mackinac Island.
Adjutant General Record gives his name as "John B.S. Shomin". Where the "S" in B.S. came from is a mystery. All other records just say "John B. Shomin". One more note, the John Shomin who moved to Harbor Springs often signed his name "John O. Shomin".

THERE ARE THREE GRAVESTONES FOR TWO JOHN SHOMINS

1. **Mackinac Island / St. Anne Cemetery. Gravestone says "John Shomin".**
   Since John B. Shomin died on Mackinac Island, this must be his grave.
2. **Harbor Springs / Lakeview Cemetery. Gravestone says "John Shomin".**
   Since John O. Shomin died in Harbor Springs, this must be his grave.
3. **Harbor Springs / Lakeview Cemetery. Gravestone says "John B. Shomin".**
   Years ago this gravestone was six feet away from the "John Shomin" gravestone.

The cemetery figured out that two stones had been ordered for same person, and both stones sat on one grave. Today the two stones stand back-to-back on top of the grave of John O. Shomin.

 All Rights Reserved by Chris Czopek

JOHN O. SHOMIN 118

Shomin, John, LaCroix. Enlisted in company K, First Sharpshooters, June 11, 1863, at LaCroix, for 3 years, age 18. Mustered July 20, 1863. Discharged at Chester, Pa., June 21, 1865. Died July 3, 1895. Buried at Harbor Springs, Mich.

Shomin, John O. – PRIME NAME, from 1894 Veteran Census
Shomin, John – Adjutant General Record name
Shomin, John Baptiste – Baptismal Name (see note below)

RESIDENCE IN 1863: LaCroix (=Cross Village), Emmet County S-AGR
In 1890: Friendship township, Emmet County S-1890 Veteran Census

DEATH DATE: July 3, 1895 S-AGR

GRAVE: Emmet County, Harbor Springs / Lakeview Cemetery (Military Stone)
GPS Lat/Long: 45.04427, -84.99899

PENSION FILE: Yes, see Appendix A
1883 Pension Roll: Yes, Emmet Co. – Harbor Springs (gsw left forearm)

VETERAN CENSUS: 1888 Yes / 1890 Yes / 1894 Yes

RAY HEREK's BOOK: Pages: Preface xii, 22, 367, 399, 435(Roster)

SPECIAL INFORMATION:
Wounded in action during the siege of Petersburg – Gun shot wound in left arm.
Company K had two men named "John Shomin"; they were cousins.
Both cousins were baptized "John Baptiste Shomin". S-ccn118a
Army records named one "John", the other "John B." S-AGR
After the war, John lived in Harbor Springs, and John B. moved to Mackinac Island.
John Shomin of Harbor Springs did not use a middle name, except for once or twice, and when he did, he used the middle initial "O".
Company K had three men named SHOMIN.
John B. Shomin and Louis Shomin were brothers. John O. Shomin was their cousin.
All three grew up together in Cross Village. All three survived the war.

NOTE TO HISTORIANS:
Shomin means "grape" (as in wild grapes). It is a common family name among the Anishinabe people, the same way "Smith" and "Brown" are common names.

 All Rights Reserved by Chris Czopek

LOUIS SHOMIN 119

Shomin, Louis, LaCroix. Enlisted in company K, First Sharpshooters, June 11, 1863, at LaCroix, for 3 years, age 19. Mustered July 20, 1863. Mustered out at Delaney House, D. C., July 28, 1865.

Shomin, Louis – PRIME NAME, from Adjutant General Record
Shornin, Louis – Name on Pension File index card
Shouman, Louis – Gravestone name
Shonin, Lewis – 1888 Veteran Census

RESIDENCE IN 1863: LaCroix (=Cross Village), Emmet County S-AGR

DEATH DATE: April 15, 1893 S-pf-119

GRAVE: Leelanau Co., Peshawbestown / Kateri Tekakwitha Catholic Cemetery
(Look for the Louis Shomin grave in the back row of the cemetery, under the trees. Military stone. Name on gravestone is "Louis Shouman". *In past years, this cemetery was named Beechwood Cemetery. It also has been called Immaculate Conception Cemetery.*)

PENSION FILE: Yes, see Appendix A
1883 Pension Roll: No

VETERAN CENSUS: 1888 Yes / 1890 Yes / 1894 Yes

RAY HEREK's BOOK: Pages: 22, 405, 435(Roster)

SPECIAL INFORMATION:
THE SHOMIN FAMILY IN COMPANY K: S-pf-119, and S-AGR
There were three soldiers from the Shomin family in Company K. All were from Cross Village. Louis Shomin and John B. Shomin were brothers. A third man, John O. Shomin, was their cousin. Louis was a farmer; he lived most of his life in Emmet County, Readmond township.
His wife was named Rosa. When he died, she received a widow's pension.
"All the Shomins are of the better class, and better educated" said one witness when he gave testimony for a pension application.
During the siege of Petersburg a shell exploded 15 feet in front of Louis Shomin. It threw dirt into his face. Many years after the war, he lost the sight in his right eye and blamed it on the exploding shell. S-pf-119
Louis was member #17 of the Washington G.A.R. Post No. 106, Cross Village, Emmet Co. He served as OG, "Officer of the Guard".

 All Rights Reserved by Chris Czopek

THOMAS SMITH 120

Smith, Thomas, Isabella. Enlisted in company K, First Sharpshooters, May 18, 1863, at Isabella, for 3 years, age 23. Mustered May 26, 1863. Sergeant March 1, 1865. Discharged at Detroit, Mich., Aug. 11, 1865.

Smith, Thomas – PRIME NAME
Shay-Go-Nay-Be, Thomas – Obituary, aka name

RESIDENCE IN 1863: Isabella township, Isabella County S-AGR
In 1890: Rosebush, Isabella township, Isabella County S-1890 Veteran Census
In 1894: Isabella township, Isabella County S-1894 Veteran Census
DEATH DATE: April 22, 1909 S-obituary, Flint Daily Journal, May 1, 1909
GRAVE: Bissing cemetery, aka Nippissing Indian Cemetery (Small gravestone)
This cemetery is located near the corner of Rosebush Road & North Isabella Road in Isabella township, Isabella County S-State Death Record
PENSION FILE: Yes, see Appendix A
1883 Pension Roll: No
VETERAN CENSUS: 1888 Yes / 1890 Yes / 1894 Yes
RAY HEREK's BOOK: Pages: 281, 435(Roster)

SPECIAL INFORMATION: On the reservation, everyone called him Thomas Shay-Go-Nay-Be. When he was a boy, he received a grammar school education. Afterwards he became a prominent member of the tribe. His influence was the reason so many men from Isabella joined Company K.

The story of Thomas Smith is without parallel in the annals of Civil War history!
The man was born without a right hand, yet he could load and fire a Springfield musket as fast as a man with two good hands! And his marksmanship was superb.
After he was mustered in as a soldier in Company K, his talent for leadership became apparent. By the end of the war Thomas Smith was the 1st Sergeant of Company K, a position of great responsibility. Tragically, during his last weeks in the army, a severe case of Erysipelas affected his eyes, and he went blind. Two or three hospitals treated him, but it was no use. He was one of the last soldiers in Company K to be discharged from the army. Back home on the Chippewa reservation in Isabella County, he was known for the rest of his life as "Blind Tom".
He received a pension from the government for his disability – after a twenty year struggle!
His obituary says he was a Minister (North Branch Methodist Episcopal Church, Isabella twp.).
Photograph located! While at Harewood Hospital in Washington D.C., the medical case of Thomas Smith was recorded, and a photograph taken, for the Army Medical Museum. The photo is now at the Otis Archives of the National Museum of Health and Medicine, Silver Spring, MD. It is labeled "1102 Continuous Photo". His shirt is off, and his right arm (missing a hand) is shown.

The book "Diba Jimooyung – A History of the Saginaw Ojibwe Anishinabek", says on Pg. 81: " [Joseph] Bradley, along with two other exceptionally competent Anishinabek, 'Blindtown' [Thomas] Smith and Phillip Gruett, could read and write and understood the details of land ownership under the U.S. system." The man named Blindtown Smith is "Blind Tom Smith", the Company K soldier.

PETER SOUTH 121

South, Peter, Bear River. Enlisted in company K, First Sharpshooters, June 17, 1863, at Bear River, for 3 years, age 18. Mustered July 20, 1863. Taken prisoner at Petersburg, Va., June 17, 1864. Died in prison at Andersonville, Ga., Dec. 19, 1864. Buried in National Cemetery at Andersonville, Ga. Grave No. 12310.

South, Peter – PRIME NAME, all spellings agree.
Pay-wau-su-quot, Peter – His true name (see note below) S-pf-121

RESIDENCE IN 1863: Bear River (=Petoskey), Emmet County S-AGR

DEATH DATE: December 19, 1864 (*Died while a POW at Andersonville Prison*) S-AGR

GRAVE: Georgia / Andersonville National Cemetery, Grave No. 12,310
GPS N32.20452 W84.13046

PENSION FILE: Yes, see Appendix A; 1883 Pension Roll: No

VETERAN CENSUS: 1888 No / 1890 No / 1894 No

RAY HEREK's BOOK: Pages: 351, 399, 414, 435(Roster)

SPECIAL INFORMATION:
He enlisted in Company K at Petoskey on June 17, 1863. Exactly one year later he was captured and sent to the infamous Andersonville prison, where he died from starvation and disease.

HOW PETER SOUTH RECEIVED HIS NAME:
His Anishinabe name was Pay-wau-su-quot. He was given the name South by a teacher at the mission school. To help his mother receive a pension, the teacher gave this testimony: "I, as mission teacher for the Presbyterian Board of Foreign Missions, named her son South as an interpretation of one of his mother's names … Shawananokwe".
[Shawananokwe = Woman from the South] S-pf-121 (Testimony of Albert Porter)

NOTE TO HISTORIANS:
Where did this "Woman From The South" come from?
Peter's mother may have been a Potawatomi from southern Michigan, who moved up north when her tribe lost their traditional lands. But this is just a guess, not a proven fact.
Further research needed.

 All Rights Reserved by Chris Czopek

GEORGE STONEMAN 122

Stoneman, George, Pentwater. Enlisted in company K, First Sharpshooters, July 4, 1863, at Pentwater, for 3 years, age 22. Mustered July 11, 1863. Wounded in action Sept., 1864. Discharged May 18, 1865.

Stoneman, George – PRIME NAME, all sources agree on spelling

RESIDENCE IN 1863: Pentwater, Oceana County S-AGR
In 1883: Hart, Oceana County S-1883 Pension Roll

DEATH DATE: July 9,1887 *(Died near Whitehall, Muskegon County)* S-pf-122
Obituary on front page of Whitehall Forum, July 14, 1887

GRAVE: Muskegon County, Whitehall /Oakhurst Cemetery S-Shier pg 377
(Buried in an unmarked grave. Oakhurst is the city cemetery for Whitehall, located at the corner of Colby Road and Peterson Road)

PENSION FILE: Yes, see Appendix A
1883 Pension Roll: Yes, Oceana Co. – Hart (No injury stated)

VETERAN CENSUS: 1888 No / 1890 No / 1894 No

RAY HEREK's BOOK: Pages: 388, 435(Roster)

SPECIAL INFORMATION:
Wounded in action at Battle of Ni River May 9-10. Lost two fingers. S-Herek pg 388
Killed in a train accident in 1887. Story on front page of Whitehall Forum, July 14, 1887. The newspaper story says "Stoneman was remarkably bright for an Indian, could speak English fluently and was the head of the band of Indians to which he belonged".

ERROR IN THE RECORDS
Adjutant General Record says he was wounded in September 1864.
Herek's book, pg 388, says he was wounded May 9-10, 1864.
Of course, it is possible he was wounded twice.

FREEMAN SUTTON 123

Sutton, Freeman. Enlisted in company K, First Sharpshooters, Feb. 10, 1865, at Grand Rapids, for 3 years, age 32. Mustered Feb. 10, 1865. Discharged at Alexandria, Va., June 9, 1865.

Sutton, Freeman – PRIME NAME, all sources agree on spelling

RESIDENCE IN 1865: Pere Marquette (=Ludington), Mason County S-Descriptive Roll, State Archives
In 1883: Custer, Mason County S-1883 Pension Roll

DEATH DATE: December 13,1890
(*Died in an accident, run over by a train.* S-ccn123a)

GRAVE: Location Unknown *(Since he lived and died near Freesoil, Mason County, he must be buried in that vicinity)*

PENSION FILE: Yes, see Appendix A
1883 Pension Roll: Yes, Mason Co. – Custer (gsw left thigh – *see note below*)

VETERAN CENSUS: 1888 No / 1890 No / 1894 No

RAY HEREK's BOOK: Pages: 407, 435(Roster)

SPECIAL INFORMATION:
During the last few months of the Civil War, the army recruited "replacements" for Company K by telling them they only need to sign up for one year, instead of three years. Freeman Sutton enlisted under those terms. Soon as he signed the papers, they sent him straight into battle!
It was the last days of the siege of Petersburg, Va. Every new soldier the Union could find was sent there to help General Grant break through the lines. The last big battle at Petersburg was on April 2, 1865. In this fight, Freeman Sutton was seriously wounded in his left leg by pieces from an exploding shell. While in the hospital, Richmond fell and Lee surrendered. The war was over!
As soon as he was well enough to walk, the army gave him his discharge and sent him home. But his leg never healed right. He was lame for the rest of his life. S-ccn123b
He collected a disability pension from the Government till his tragic death in 1890. (*Died in an accident, run over by a train.* S-ccn123a)

NOTE TO HISTORIANS REGARDING HIS WOUND:
The abbreviation "gsw" means Gun Shot Wound. It usually means injury from a rifle bullet. However, a wound from an exploding artillery shell was also classified as a Gun Shot Wound.

FRANCIS TABASASH 124

Tabasasch, Francis, LaCroix. Enlisted in company K, First Sharpshooters, as Corporal, June 11, 1863, at LaCroix, for 3 years, age 38. Mustered July 20, 1863. Promoted Sergeant. Mustered out at Delaney House, D. C., July 28, 1865.

Tabasash, Francis – PRIME NAME, G.A.R. Post No. 106 records
Tabasasch, Francis – Adjutant General Record name
Tabash, Francis – 1888 & 1894 Veteran Census name
Assinnaby, Francis – 1860 US Census (see note below)

RESIDENCE IN 1863: LaCroix (=Cross Village), Emmet County S-AGR
After the war, he returned to Cross Village and lived there the rest of his life

DEATH DATE: November 22, 1912 S-State Death Record
GRAVE: Emmet County, Cross Village / Holy Cross Cemetery (Military stone)
(*Location: Holy Cross Cemetery is on Levering road, exactly one mile east of Cross Village*)

PENSION FILE: Yes, see Appendix A; 1883 Pension Roll: No
VETERAN CENSUS: 1888 Yes / 1890 Yes / 1894 Yes

RAY HEREK's BOOK: Pages: 36, 281, 372, 435(Roster)

SPECIAL INFORMATION:
Enlisted at age 38 as Corporal. Later was promoted to First Sergeant of Company K. S-Herek pg 281
Francis was a charter member of the Washington G.A.R. Post No. 106, Cross Village, Emmet Co. He was never the Post Commander, but he was JVC (Junior Vice Commander) which is third from the top in the chain of command.
On the internet is a photograph taken around 1890 of Francis Tabasash /Tabasasch wearing a G.A.R. ribbon and sitting in front of his house with a rifle across his lap. The tall hat he is wearing is called a "Hardee Hat", and it was issued to soldiers in the Civil War. S-ccn124a
There is a quilt at the MSU Textile Museum in Lansing made by his daughter, Madeline.

NOTE TO HISTORIANS:
Before the war, his name was Francis Assinnaby. S-1860 US Census
At the time he enlisted in Company K, he changed his last name to Tabasash and used that for the rest of his life.
There is a controversy over the correct spelling of his name. Some records say "Tabasash", others say "Tabasasch". Since every census and civilian record says Tabasash, and so does the G.A.R. Post roster, and the only records that say Tabasasch with a "c" are army records and his gravestone, I believe an army clerk misspelled his name when he enlisted, and the mistake was repeated over and over. Based on that, I say Tabasash is the correct spelling.

ANTOINE TABYANT 125

Tab-yant, Antoine, Little Traverse. Enlisted in company K, First Sharpshooters, June 14, 1863, at Little Traverse, for 3 years, age 35. Mustered July 20, 1863. Wounded in action Sept., 1864. Mustered out at Delaney House, D. C., July 28, 1865. Died May 22, 1894. Buried at Harbor Springs, Mich.

Tabyant, Antoine – PRIME NAME, from Pension File
Tab-yant, Antoine – Adjutant General Record name
Tobyant, Jno. – Gravestone name
Tabiah, Antoine – 1890 Veteran Census
Tobiah, Antoine – 1888 Veteran Census
Tab Zout, Antoine – 1883 Pension Roll

RESIDENCE IN 1863: Little Traverse (=Harbor Springs), Emmet County S-AGR
In 1890: Friendship township, Emmet County S-1890 Veteran Census

DEATH DATE: May 22, 1894 S-AGR

GRAVE: Emmet County, Harbor Springs / Lakeview Cemetery (Military Stone)
Name on gravestone says "Jno. Tobyant" (Jno = Johnathan or John)

PENSION FILE: Yes, see Appendix A
1883 Pension Roll: Yes, Emmet Co. – Harbor Springs (gsw right shoulder).
Note: gsw = gun shot wound

VETERAN CENSUS: 1888 Yes / 1890 Yes / 1894 No

RAY HEREK's BOOK: Pages: Preface xii, 367, 435(Roster)

SPECIAL INFORMATION:
"Antoine Tabyant is an honest trustworthy Indian, one of the best I ever knew"
This comment appears in the pension file papers for Company K soldier John Jacko, and shows Antoine was a respected member of the Harbor Springs community.
Wounded in battle. The 1883 Pension Roll says he received a gun shot wound in right shoulder, but the 1890 Veteran Census says he was wounded in the neck on the right side.
The Crooked Tree, by John C. Wright, has an anecdotal story about Antoine Tabyant.
It describes what happened when Antoine returned home from the Civil War.
(Note, this story appears in the 1917 edition of the book. It is missing from the 1915 edition).

JOSEPH TAZHEDEWIN 126

Ta-ohe-de-niw, Joseph, Pentwater. Enlisted in company K, First Sharpshooters, July 4, 1863, at Elbridge, for 3 years, age 19. Mustered July 11, 1863. Wounded in action Sept., 1864. Discharged for disability at Alexandria, Va., Aug. 18, 1864.

Ta-zhe-de-win, Joseph – PRIME NAME, from Pension File
Ta-ohe-de-niw, Joseph – Adjutant General Record name
Tashedowin, Joseph – Adjutant General Annual Report
Toshidowe, Joseph – 1890 Veteran Census / Index book
Poneshing, Joseph – County Death Record
Ponishing, Joe – Newspaper obituary Oceana Herald, January 12, 1923
Po-ne'-sing – Name on photo at Grand Rapids Public Museum

RESIDENCE IN 1863: Pentwater, Oceana County S-AGR
After the war: He lived in Elbridge township, Oceana County, but received his mail at the Stetson (=Walkerville) Post Office, Levitt township, Oceana County S-pf-126
DEATH DATE: January 3, 1923 S-County Death Record
GRAVE: Oceana County, Elbridge twp. / St. Joseph Catholic Cemetery S-obituary
(No gravestone and no burial record exists, so location of his grave in the cemetery is unknown)
PENSION FILE: Yes, see Appendix A
1883 Pension Roll: Yes, Oceana Co. – Hart (loss of little left finger)
VETERAN CENSUS: 1888 Yes / 1890 Yes / 1894 Yes
RAY HEREK's BOOK: Pages: 410, 435(Roster)

SPECIAL INFORMATION:
JOSEPH WAS A MAN WITH TWO NAMES. He used **Joseph Tazhedewin** when he joined the army, and for the 1890 Veteran Census, and for his army pension paperwork.
He used **Joseph Poneshing** as his family name & legal name. Poneshing was also used for his obituary and County Death Record. His children used variations of this name as their last name. Grandson Henry J. Pawneshing was a highly decorated Marine in WWII and Korea. S-ccn126a
The Grand Rapids Public Museum has on display a photograph of Joseph Poneshing taken in 1914, showing him sitting on the front porch of his home surrounded by his family. The picture came from the Grand Rapids Public Library's Local History Collection. S-ccn126b
ERROR IN THE RECORDS: Notice the Adjutant General Record says this man was wounded in September, and discharged from the army in August. Can you see the error?
A finger on his left hand was shot off at Petersburg. He was discharged August 18, 1864 as a disabled soldier. He later received a disability pension from the government.

NOTE TO HISTORIANS: When researching this man, you will run across a town named Stetson. Stetson is the old name for Walkerville.

MOSES THOMAS 127

Moses, Thomas. Enlisted in company K, First Sharpshooters, Jan. 27, 1865, at Detroit, for 3 years, age 18. Mustered Jan. 28, 1865. Mustered out at Delaney House, D. C., July 28, 1865.

Thomas, Moses – PRIME NAME, from Pension File
Moses, Thomas – Adjutant General Record name (incorrect name)

RESIDENCES:
1847: Born in Bay City, Michigan, in 1847 S-Ontario Province Death Certificate
1865: White Pigeon, St. Joseph County S-Descriptive Roll, State Archives
1880: Pinconning, Bay County, Michigan S-1880 US Census
1883-1933: Kettle Point, Canada (Indian Reserve No. 44) S-Ontario Province Death Certificate
(For the location of Kettle Point, see note below)

DEATH DATE: August 25, 1933 S-pf-127
(Died in Canada, Ontario, Lambton County, on the reservation at Kettle Point (Reserve No. 44)
GRAVE: Location Unknown *(Buried on the reservation at Kettle Point, Canada)*

PENSION FILE: Yes, see Appendix A
1883 Pension Roll: No

VETERAN CENSUS: 1888 No / 1890 No / 1894 No

RAY HEREK's BOOK: Pages: 435(Roster), 496

SPECIAL INFORMATION:
Moses Thomas died in 1933 - the last surviving member of Company K.
He was born in Bay City and later lived in Pinconning. These facts indicate he was a Chippewa. Around 1883 he moved to Canada and lived there on an Indian reservation (Kettle Point, Ontario). A photograph exists of him sitting outdoors behind his house with his wife Julia. S-ccn127a
Warning! There is a major error in the official records – The Adjutant General Record says his last name is Moses, but this is wrong. **His name is Moses Thomas, not Thomas Moses.** S-pf-127
In 1865 he enlisted in the army in Detroit and said his home was White Pigeon, St. Joseph County S-Descriptive Roll, State Archives. After the war, he lived in Pinconning, MI and was recorded on the 1880 US Census. In 1883 he moved to Kettle Point, Canada. S-Ontario Province Death Certificate

NOTE TO HISTORIANS:
Kettle Point is located north of Sarnia, Ontario, on the Lake Huron shoreline. You can find it on a map near the town of Forest. It is in Lambton County, Ontario.
Most of the land on Kettle Point is an Indian reservation known as "Indian reserve 44".
The tribe living there is officially named "Chippewas of Kettle and Stony Point First Nation".

ROBERT VALENTINE 128

Valentine, Robert. Enlisted in company K, First Sharpshooters, Aug. 24, 1864, at Jackson, for 3 years, age 23. Mustered Aug. 24, 1864. Substitute for Joseph Mills. Discharged at Detroit, Mich., Aug. 11, 1865.

Valentine, Robert – PRIME NAME, from Adjutant General Record
Valintine, Robert – Pension File name

RESIDENCE IN 1864: Recruited in the Upper Peninsula S-Herek pg 250
Army records say he enlisted in Jackson, Michigan S-AGR
Credited to Lansing for draft quota purposes S-Descriptive Roll, State Archives
1866: Living in Odanah, Wisconsin S-Shier pg 400

DEATH DATE: December 22, 1871

GRAVE: Location Unknown *(Possibly buried on the Bad River Reservation of the Lake Superior Tribe of Chippewa, Odanah, Wisconsin)*

PENSION FILE: Yes, see Appendix A; 1883 Pension Roll: No

VETERAN CENSUS: 1888 No / 1890 No / 1894 No

RAY HEREK's BOOK: Pages: 250, 435(Roster)

SPECIAL INFORMATION:
Robert Valentine enlisted in Jackson, Michigan, as a substitute for Joseph Mills.
It was common for a man who was being drafted to pay someone to take his place.
Robert Valentine probably received a hundred dollars for doing this, big money in those days.
The reason why Lansing receives a draft-credit for Valentine is a matter for speculation.
Herek's book "These Men Have Seen Hard Service" describes Valentine as a professional hunter recruited in the Upper Peninsula. S-Herek pg 251
Most of the sharpshooters were discharged in Washington, D.C. on July 28. The fact that Robert Valentine was discharged in Detroit on August 11 suggests he was seriously ill in a hospital, and his discharge was delayed till he was well enough to travel home.
First thing he did when he arrived home was get married! Discharged Aug 11, married Aug 27. In Michigan's UP, Baraga County, at the Methodist Mission at L'Anse, he married the sister of his army friend Henry Wassagezic aka Henry Condecon. Her name was Mary Condecon, and she was the daughter of Chief Waasegiizhig. They were married August 27, 1865. S-Shier pg 400
There is no record of children. Robert Valentine died December 22, 1871, at the young age of 30. Cause of death, unknown. His widow, Mary, lived in Odanah till 1925. S-Shier pg 401

THOMAS WABANO 129

Wabano, Thomas, Isabella. Enlisted in company K, First Sharpshooters, May 18, 1863, at Isabella, for 3 years, age 24. Mustered May 26, 1863. Died of disease at Camp Douglas, Chicago, Ill., Jan. 7, 1864.

Wabano, Thomas – PRIME NAME, from Adjutant General Record
Wau-bau-no, Thomas – Pension File name
Wa-Bu-No, Thos. – Gravestone name and G.A.R. Post name
Waubanoo, Thomas – Ray Herek's book uses this spelling
Wabesis, Thomas – Adjutant General Annual Report

RESIDENCE IN 1863: Isabella township, Isabella County S-AGR

DEATH DATE: January 7, 1864 S-AGR
(*Note: he died of disease in Isabella County, not at Camp Douglas, Chicago - see note below*)

GRAVE: Isabella Co., Mt. Pleasant / Riverside Cemetery (Military Gravestone)
(His gravestone near the big flagpole & stack of cannon balls is a memorial stone, the body of Thomas is buried elsewhere, says Bob Grove, Graves Officer for the Sons of Union Veterans)

PENSION FILE: Yes, see Appendix A
1883 Pension Roll: No

VETERAN CENSUS: 1888 No / 1890 No / 1894 No

RAY HEREK's BOOK: Pages: 419, 435(Roster)

SPECIAL INFORMATION:
When he joined the army, he left behind a wife and newborn baby daughter. S-pf-129
While guarding prisoners at Camp Douglas in Chicago, Illinois, Thomas Wabuno became seriously ill. He was sent home to Michigan on sick furlough. Three days later, he died in the cabin of his father. S-pf-129
The Mt. Pleasant G.A.R. was named WA-BU-NO Post No. 250 (Active 1884 – 1923).
It was named after Thomas Wabano of Company K, 1st Michigan Sharpshooters.
It was the only G.A.R. Post in the country named after a Native American.

NOTE TO HISTORIANS:
His pension file gives five different dates for his death:
* Dec 31,1862 * Jan 7, 1863 * Jan 7, 1864 * Dec 31, 1864 * Jan 15, 1865*
The Adjutant General Record says he died Jan 7, 1864, and that is the date I accept as correct.
ERROR IN THE RECORDS: Adj. General Record says he died at Camp Douglas, Chicago. Not true! His pension file says he was sent home on furlough and died in Isabella County.

CHARLES WABESIS 130

Waubesis, Charles, Pentwater. Enlisted in company K, First Sharpshooters, as Sergeant, July 4, 1863, at Elbridge, for 3 years, age 25. Mustered July 11, 1863. Retained by civil authority at Chicago, Ill., for murder. Sentenced to imprisonment for life.

Wabesis, Charles – PRIME NAME, from Pension File
Waubesis, Charles – Adjutant General Record name
Wabsis, Charles – 1890 Veteran Census
"Charles Naw-we-caw Wah-be-sis" – Find-A-Grave database

RESIDENCE IN 1863: Pentwater, Oceana County S-AGR
In 1891: Elbridge township, Oceana County S-pf-130
DEATH DATE: December 21, 1891 S-pf-130
GRAVE: Oceana County, Elbridge twp. / St. Joseph Catholic Cemetery (Military Stone) *This cemetery is located near the intersection of 144th and Clock road Specifically, on the east side of 144th, between Clock rd. and Harrison rd.*
PENSION FILE: Yes, see Appendix A; 1883 Pension Roll: No
VETERAN CENSUS: 1888 No / 1890 Yes / 1894 No
RAY HEREK's BOOK: Pages: 192, 435(Roster), 462

SPECIAL INFORMATION:
Charles Wabesis and his younger brother John joined the army together on the 4th of July. The Adjutant General Record says Charles Wabesis committed murder and was sentenced to life in prison. **But his pension file tells a different story**. When his case came to trial, he received a 6 month sentence for "Getting mixed up in a row". After serving his sentence, he went home to Elbridge without reporting to the army. For this reason, he was charged with being a deserter. Many years later he explained to the War Department what had happened, and the charge of desertion was removed from his record**. He was given an honorable discharge** dated July 28, 1865, which was the same day the rest of Company K was discharged.
Controversy Over Death Date: His pension file says he died December 21, 1891. However, the Death Records in the office of the Oceana County Clerk, Liber 1, Pg 138, say he died December 23, 1891. Which date is correct? Further research needed.

NOTES FOR FURTHER RESEARCH:
You can read the story about the incident, a fight in a Chicago saloon where one man was stabbed, in the Chicago Tribune, March 19, 1864, page 4, article titled THE MOYNIHAN MURDER. The criminal court records of this case were destroyed in a fire October 9, 1871. S-pf-130
Suggestion: Check the correspondence of the governor of Illinois Richard Yates (1861-1865) and Richard Oglesby (Jan 1865 to 1869) for a letter from Colonel Charles DeLand of the 1st Michigan Sharpshooters. I have been told Colonel DeLand asked the Governor to pardon Charles Wabesis. This letter, if found, may give new information about what happened in the saloon.

JOHN WABESIS 131

Wa-be-sis, John, Pentwater. Enlisted in company K, First Sharpshooters, July 4, 1863, at Elbridge, for 3 years, age 23. Mustered July 11, 1863. Missing in action at Petersburg, Va., July 30, 1864.

Wabesis, John – PRIME NAME, from Pension File
Wa-be-sis, John – Adjutant General Record name

RESIDENCE IN 1863: Elbridge township, Oceana County S-AGR

DEATH DATE: July 30, 1864 (*KIA at Battle of the Crater, Petersburg, VA*) S-AGR
(Note: KIA = Killed In Action)

GRAVE: Location Unknown

PENSION FILE: Yes, see Appendix A; 1883 Pension Roll: No

VETERAN CENSUS: 1888 No / 1890 No / 1894 No

RAY HEREK's BOOK: Pages: 403, 435(Roster)

SPECIAL INFORMATION:
John Waw-be-sis was killed in the line of his duty by rebels and enemys of the government. These blunt words were spoken by Pvt. Josiah Light, a fellow soldier in Company K. S-pf-131
The Adjutant General records said John was "Missing in Action". This caused a problem after the war for the widow of John Wabesis, when she tried to get a pension on behalf of her two children. The survivors of Company K rallied to her aid. What she needed was proof that **her husband had been killed in the fighting on July 30, which history books call "The Battle of the Crater".** Sergeant Antoine Scott had proved himself a hero in this fight – was recommended for the Medal of Honor. Sergeant Scott helped the widow by sending this testimony to the government:
"John Wabesis and I was in a fort [the crater] … a shell dropped down and bursted and killed several, and I saw John Wabesis lying on the ground … lifeless and I suppose he was dead. I know these facts from personal observation … signed Antoine Scott". S-pf-131
Other veterans of Company K also gave testimony. The request for a pension was approved!

There is a Veterans Monument in Hart, Michigan (Oceana County) with his name on it.
Married March 15, 1856 in Newaygo Co. on the bank of Tamarack Creek to Mohtwaynekenoh.
One baby girl survived to adulthood, named Kahbaynoquay, born Oct 15, 1857. S-Shier pg 416
John Wabesis and his older brother Charles joined the army together on the 4th of July.

NOTES FOR HISTORIANS:
There is a mystery soldier in Company K named John Pakemaboga, killed in action on July 30. Could John Pakemaboga & John Wabesis be the same person? More research needed.

 All Rights Reserved by Chris Czopek

JOSEPH WAKAZOO 132

Wakazoo, or Wau-ka-zoo, Joseph, Grand Traverse. Enlisted in company H, Sixteenth Infantry, Nov. 8, 1861, at Hall's Hill, Va., age 23. Mustered Jan. 14, 1862. Missing in action Aug. 27, 1862. Transferred to company K, First U. S. Sharpshooters, Nov. 27, 1863. Discharged at Detroit, Mich., Feb. 15, 1865.

Wakazoo, Joseph – PRIME NAME, Adjutant General Record and Pension File
Waukazoo, Joseph – Adjutant General Record, aka name

RESIDENCE IN 1861: Grand Traverse Bay (=Leelanau County) S-Descriptive Roll
From 1865-1910 he lived in Minnesota S-ccn132a
DEATH DATE: July 18, 1910 S-pf-132
GRAVE: Location Unknown
(*He died on the White Earth Reservation, Mahnomen Co., in Northwestern Minnesota* S-ccn132b)
PENSION FILE: Yes, see Appendix A; 1883 Pension Roll: No
VETERAN CENSUS: 1888 No / 1890 No / 1894 No
RAY HEREK's BOOK: This soldier is not mentioned

SPECIAL INFORMATION:
Chief Peter Waukazoo is remembered for moving his tribe from Holland to the Leelanau Peninsula. **Joseph Wakazoo was the Chief's son.** Before the war, Joe received an education in Ohio at the Twinsburg Institute in Summit County, Ohio. Among his classmates were Andrew Jackson Blackbird and Simon Pokagon. S-ccn132c
When the Civil War began, he enlisted in the 16th Michigan Infantry, Co. H.
He was transferred by Order of the War Dept. to Company K, 1st Mich. Sharpshooters. S-ccn132d
No doubt this seasoned soldier received a warm welcome from his first cousin Payson Wolf and the other soldiers from the Leelanau area. **Wounded in action at the Battle of Spotsylvania on May 12, 1864** (shot through the left hand. Spent the rest of the war in hospitals). S-pf-132
When the war was over, Joseph decided to become a minister. He spent the rest of his life working as a missionary among the Ojibwe tribes in Minnesota.
There is a picture of Joseph Wakazoo in a missionary magazine published in 1885. S-ccn132e
The Rev. Joseph Wakazoo died in Minnesota on the White Earth Reservation in 1910, at age 84.
A good source for information about Joseph Wakazoo is the book Old Wing Mission, by Robert Swierenga and William Van Appledorn, Published 2008 by Eerdmans Publishing, Grand Rapids.

NOTES FOR HISTORIANS:
His name is spelled two ways: Wakazoo / Waukazoo. I choose **Wakazoo** because that is how Joseph signed his name. But his family also uses Waukazoo or Waukazo as the spelling.
He appears on the Taggart Roll as "Jos. Wakazoo, Age 60, Residence: Lengby, Polk Co., Minn.".
ERROR IN THE RECORDS: Due to a clerical error, Joseph was listed as transferring to the 1st US Sharpshooters, when it should have said 1st Michigan Sharpshooters.
Because of this error, he is not on the official roster of Company K in the Adjutant General Record. Consequently, his name is missing on many rosters of Company K.

HENRY WASSAGEZIC 133

Wassagezic, Henry. Enlisted in company K, First Sharpshooters, Sept. 24, 1864, at Pewabic, for 3 years, age 17. Mustered Oct. 20, 1864. Joined regiment at Peebles Farm, Nov. 4, 1864. Mustered out at Delaney House, D. C., July 28, 1865.

Wassagezic, Henry – PRIME NAME, Adjutant General Record
Wassagezie, Henry – Pension File
Condeacon, Henry – Gravestone Name
Nawzhekegezhig (Clouds Disappearing) – Name on 1899 photograph

RESIDENCE IN 1864: Pewabic, Houghton County, Michigan S-AGR
(In 1864, Pewabic was part of Ontonagon County, Michigan)
In the latter half of his life, he lived on the Bad River reservation at Odanah, in Ashland County, Wisconsin S-pf-133 (pension file for Henry Wassagezic)
DEATH DATE: April 16, 1922 S-pf-133 (pension file for Henry Wassagezic)
GRAVE: Wisconsin, Ashland County, Odanah / Bad River Cemetery (Military Gravestone) *Buried on the Bad River Reservation of the Lake Superior Tribe of Chippewa*
PENSION FILE: Yes, see Appendix A; 1883 Pension Roll: No
VETERAN CENSUS: 1888 No / 1890 No / 1894 No
RAY HEREK's BOOK: Pages: 435(Roster)

SPECIAL INFORMATION:
Another soldier in Company K, James Arwonogezice, sometimes used "Wasagesik" as his last name. Be careful not to confuse Henry Wassagezic with James Wasagesik.
Henry used two names: When he joined the army, he was "Henry Wassagezic". After the war, he used the name "Henry Condecon / Condeacon". He was a member of the Lake Superior Chippewa and always lived either in the Upper Peninsula of Michigan or on the Bad River Reservation in Wisconsin. Henry was born April 30, 1847 at the Bad River Falls near Odanah, Wisconsin. When he joined the army in 1864, he was living in Pewabic, a town on Michigan's Keweenaw peninsula. S-pf-133 Henry came from a prominent family – His grandfather was Okondekon (aka Konteka), Chief of the Ontonagon Band of Chippewa, who signed the 1842 & 1854 Treaties of La Pointe; the Chief is best remembered for selling the famous Ontonagon Copper Boulder that ended up at the Smithsonian Institute Museum in Washington S-ccn133a. Henry's father was Chief Wassageshik Okondekon, who made an epic trip to Washington in 1848 to meet the President, and who also signed the 1854 Treaty of La Pointe S-ccn133b.
This Civil War veteran was an active member of the Bad River tribe in Wisconsin, attending tribal meetings at the Council House in Odanah and serving on committees S-ccn133c.
Married twice, but no children survived (his only son Thomas Condecon Jr. died at the age of 20). Henry was a member of the George A. Custer G.A.R. Post No. 140, Ashland, Wisconsin. **There is a photograph of him wearing his G.A.R. membership badge** S-ccn133d. When he died in 1922, six members of the Ashland G.A.R. post traveled through a big snowstorm to attend his funeral on the reservation. Henry was buried with full military honors. S-ccn133b.

JAMES V. WATSON 134

Watson, James V., Pentwater. Enlisted in company K, First Sharpshooters, July 4, 1863, at Elbridge, for 3 years, age 44. Mustered July 11, 1863. Discharged for disability at Chicago, Ill., June 20, 1864.

Watson, James V. – PRIME NAME, all sources agree on spelling
Watson, James B. – 1860 US Census

RESIDENCE IN 1863: Pentwater, Oceana County S-AGR
In 1860: Cheshire township, Allegan County S-1860 US Census
In 1870: Elbridge twp., Oceana Co. S-Oceana County Death Record of daughter Alice Watson

DEATH DATE: Unknown

GRAVE: Location Unknown

PENSION FILE: No Pension File exists

VETERAN CENSUS: 1888 No / 1890 No / 1894 No

RAY HEREK's BOOK: Pages: 410, 435(Roster)

SPECIAL INFORMATION:
He was 44 years old when he joined the army. In less than a year his health broke down and the army sent him home. He never saw battle. He never applied for an invalid pension.

NOTES FOR HISTORIANS:
James Watson was born c. 1819-1820 (based on age at time of enlistment). Was living in Canada in January 1847 (his son Amos was born there). On the 1860 US Census for Allegan County, Cheshire township, he appears as James B. Watson, with a wife, Mary B. Watson, and son, Amos. James Watson was 44 years old when he enlisted in the army. His health immediately broke down. His son, Amos, received word that his father was going to die, and made a trip to Camp Douglas, Chicago to see him (S-pf-131 pension file for George W. Mogage). But the father survived. The army gave James Watson a discharge for disability and let him go home. James never applied for a pension. A death record for his two year old daughter Alice Watson indicates he was still alive in 1870 and living in Elbridge township, Oceana County. The death date for James Watson is unknown.

Amos Watson, son of James, settled on the reservation of the Nottawaseppi Huron Potawatomi near Athens. From time to time he contributed news items for the local newspaper under the name "Wah-Nuh-Wah". Amos lived to a ripe old age and died in 1926 S-State Death Record.

 All Rights Reserved by Chris Czopek

JOHN WAUBENOO 135

Wau-be-noo, John, Isabella. Enlisted in company K, First Sharpshooters, May 18, 1863, at Isabella, for 3 years, age 22. Mustered May 26, 1863. Corporal. Mustered out at Delaney House, D. C., July 28, 1865.

Wau-be-noo, John – PRIME NAME, Adjutant General Record
Waubenoo, John – Pension File name
Wanbenoo, John – Pension File, aka name
Waw-be-noo, John – 1888 Veteran Census
Waw-be-naw, John – 1890 Veteran Census
Wah-Bah-Noo, John – Spelling used by his descendants S-ccn135a
Waubenau, John – Name on State Death Record
Wabana, John – 1894 Veteran Census

RESIDENCE IN 1863: Isabella township, Isabella County S-AGR
In 1888: Van Decar, Nottawa township, Isabella County S-1888 Veteran Census
In 1894: Nottawa township, Isabella County S-1894 Veteran Census

DEATH DATE: March 25, 1910 S-State Death Record

GRAVE: Isabella County, Nottawa township / Fairview Cemetery (Military stone)
Section 1, Lot number 83. GPS N43.68403, W84.95787
(Fairview Cemetery is located south of Weidman, on the corner of Weidman Rd & Johnson Rd)

PENSION FILE: Yes, see Appendix A; 1883 Pension Roll: No

VETERAN CENSUS: 1888 Yes / 1890 Yes / 1894 Yes

RAY HEREK's BOOK: Pages: 281, 435(Roster)

SPECIAL INFORMATION:
John Waubenoo served in Company K from the beginning to the end. He saw it all.
Records indicate he fought in every battle. His rank was Corporal.
Had the good fortune to be untouched by bullets. No record of serious wounds or illness.
After the war, he went home to his wife in Isabella County. They lived on a farm near Weidman and raised a family.
When fellow Company K soldier James Ohbowakemo died, John Waubenoo and his wife took care of his baby daughter, a child named Mary (aka "little Wawsechewawnoquay").
This girl grew up and married Amos Elk on September 12, 1904. S-Shier pgs 282-283
When John retired from farming, he kept busy making baskets and selling them. S-ccn135b

 All Rights Reserved by Chris Czopek

MICHAEL WAYAUBEMIND 136

Wayaubemind, Michael, Bear River. Enlisted in company K, First Sharpshooters, June 17, 1863, at Bear River, for 3 years, age 19. Mustered July 20, 1863. Wounded in action June 17, 1864. Discharged at Washington, D. C., June 24, 1865.

Wayaubemind, Michael – PRIME NAME, Adjutant General Record
Wayohlamin, Mitchell – Name on Death Record for his son Enos Mitchell

RESIDENCE IN 1863: Bear River (=Petoskey), Emmet County S-AGR

DEATH DATE: February 7, 1881 S-Probate Records, see note below

GRAVE: Location Unknown

PENSION FILE: No Pension File exists

VETERAN CENSUS: 1888 No / 1890 No / 1894 No

RAY HEREK's BOOK: Pages: 392, 396, 435(Roster)

SPECIAL INFORMATION:
He was an Odawa from Petoskey, a member of the Bear Creek Band. S-1855 Annuity Roll
Michael Wayaubemind met Lt. Garret Graveraet on June 17, 1863 and agreed to enlist in Company K. Exactly one year later, on the same date, both men were wounded in battle and sent to hospitals in Washington D.C. S-Descriptive Roll, State Archives
Michael Wayaubemind recovered from his wounds.
After the war, he went home, married, and raised a family.
His wife wasTeresa Missinini of Burt Lake. He married her on Sept 4, 1865, in Harbor Springs.
Father Zorn conducted the service. Charles Bushaw of Company K was a witness. S-Shier pg 435
You can find Michael on the Durant Roll as Michael We-aw-be-mind (see Field Notes #9-pg30).
The best source for information about his family is found in the Probate Record for his sister, Mary Wayaubemind, Charlevoix County, 1884. It states that the father of Company K soldier Michael Wayaubemind was Joseph We-aw-be-mind (see Durant Roll Field Notes #12-pg30).

ERROR IN THE RECORDS: The army had written down his name wrong - his first name was Mitchell, not Michael. His sons used their father's first name, Mitchell, as their last name. For example, one son went by the name Enos Mitchell. S-State Death Record

NOTE TO HISTORIANS: According to genealogist Quita V. Shier, there was a special connection between Michael Wayaubemind and another soldier in Company K named Antoine Scott. The two men were half-brothers ! Same mother, different fathers.
For complete details, see her book "Warriors In Mr. Lincoln's Army", pg 332 and pg 435.

WILLIAM WAYGESHEGOING 137

Way-ge-she-going, William, Isabella. Enlisted in company K, First Sharpshooters, May 18, 1863, at Isabella, for 3 years, age 22. Mustered May 26, 1863. Discharged at Washington, D. C., June 3, 1865.

Way-ge-she-going, William – PRIME NAME, Adjutant General Record
Waygeshegoing, William – Descriptive Roll, State Archives
Waggishegomy, William – Pension File name
Way-ge-zhe-gomy – Probate papers, estate of Martha W. Collins, 1916
Gisecomey, William – 1890 Veteran Census
Wegesaquina, William – 1888 Veteran Census
Wah-sas-cum – Probate papers, estate of Martha W. Collins, 1916
Long-Bill – Probate papers, estate of Martha W. Collins, 1916

RESIDENCE IN 1863: Isabella township, Isabella County S-AGR
In 1890: Taymouth township, Saginaw County S-1890 Veteran Census
In 1897: St. Charles, Saginaw County S-Probate papers, estate of Martha W. Collins

DEATH DATE: March 27, 1897 S-Shier pg 438
Died in an accident in March 1897 when struck by a train near St. Charles, MI
See story in newspaper Chesaning Argus, April 3, 1897. S-ccn137a

GRAVE: Location Unknown *(most likely St. Charles Riverside Cemetery* S-ccn137b*)*

PENSION FILE: Yes, see Appendix A
1883 Pension Roll: No

VETERAN CENSUS: 1888 Yes / 1890 Yes / 1894 No

RAY HEREK's BOOK: Pages: 435(Roster)

SPECIAL INFORMATION:
People could not pronounce his Indian name, so they just called him "Long Bill".
Died in 1897, run over by a train. Witnesses say he was NOT intoxicated; seems he had a heart attack and fell on the tracks, dead before the train struck him. S-Chesaning Argus, April 3, 1897
His daughter, Martha W. Collins, died in 1916. On file at National Archives in Washington D.C. are the probate records of her estate from the BIA. In these records you will find detailed information on his family, his descendants, and the many names he used.
Married three times. Third wife was Julia Payshon Chatfield, married in Isabella Co. Dec 1876. Two children, Jane and Thomas. Both went to Mt. Pleasant Indian School, and both ended up living and working in Lansing. Both buried in Mt. Hope Cemetery, section Q. S-Shier pg 441

 All Rights Reserved by Chris Czopek

PETER WELLS 138

Wells, Peter, Bear River. Enlisted in company K, First Sharpshooters, June 17, 1863, at Little Traverse, for 3 years, age 18. Mustered July 20, 1863. Mustered out at Delaney House, D. C., July 28, 1865.

Wells, Peter – PRIME NAME, All sources agree on spelling

RESIDENCE IN 1863: Bear River (=Petoskey), Emmet County S-AGR
In 1866: Hayes township, Charlevoix County S-pf-138

DEATH DATE: December 6, 1888 S-County Death Record

GRAVE: Location Unknown

PENSION FILE: Yes, see Appendix A
1883 Pension Roll: No

VETERAN CENSUS: 1888 No / 1890 No / 1894 No

RAY HEREK's BOOK: Pages: 281, 435(Roster)

SPECIAL INFORMATION:
On June 17, 1863, Lt. Graveraet came to Bear River (today called Petoskey) seeking volunteers. Peter Wells answered the call and went off to war. Was in Company K from start to finish, from the first battle to the last fight. Had a charmed life, was never wounded and never seriously ill. Promoted to Corporal.

NOTE TO HISTORIANS:
The Charlevoix County Clerk's office has a County Death Record for Peter Wells. It says he died December 6, 1888, in an accident. S-County Deaths, Vol 1, pg 40 (Red ledger book)
The story of the accident was front page news in the **Charlevoix Sentinel on Dec. 12, 1888**.
The newspaper said Peter was a veteran of the 1st Michigan Engineers and Mechanics, an obvious mistake. Peter Wells served in the 1st Michigan Sharpshooters, Company K.
The pension file of Francis Tabasash has a note dated February 26, 1890 that indicates Peter Wells was alive and living in Horton Bay. Apparently, Francis was not aware that Peter had recently died.

THOMAS WESAW 139

Wesaw, Thomas. Enlisted in company K, First Sharpshooters, Sept. 24, 1864, at Van Buren, for 1 year, age 22. Mustered Sept. 24, 1864. Discharged.

Wesaw, Thomas – PRIME NAME, Adjutant General Record
Wezoo, Thomas – Pension File & Gravestone
Wezo, Thomas – 1890 Veteran Census

RESIDENCE IN 1864: Van Buren township, Wayne County S-AGR
In 1884-1886: Bradley and Martin, Allegan County S-pf-139
In 1890: Athens, Calhoun County (On the Nottawaseppi Huron Potawatomi reservation)
DEATH DATE: December 17, 1918 S-State Death Record
GRAVE: Calhoun Co., Athens township, Indian Mission Cemetery (Military stone)
(*Name on gravestone is "Thos. Wezoo". The cemetery is on the reservation of the Nottawaseppi Huron Band of Potawatomi, located south of Battle Creek near Athens.*)
PENSION FILE: Yes, see Appendix A; 1883 Pension Roll: No
VETERAN CENSUS: 1888 No / 1890 Yes / 1894 Yes
RAY HEREK's BOOK: Pages: 406, 435(Roster)

SPECIAL INFORMATION:
Wesaw is the name of a prominent Potawatomi family in Southwestern Michigan.
He enlisted in Van Buren. Although many Wesaws live in Van Buren County near Hartford, the "Van Buren" where he enlisted is a township in Wayne County. The Huron river flows through it. This river was the former home of his tribe, the Huron Nottawaseppi Potawatomi. In February of 1865 the Civil War was almost over. But there were still a few battles to be fought. Company K urgently needed "replacements" before going back into action. Recruiters visited the reservations in Michigan offering a one year enlistment. On those terms, Thomas Wesaw enlisted. He arrived at Petersburg just in time for the historic Battle of Ft. Stedman on March 25, 1865. History books call it General Lee's last offensive battle. When Company K and other Union troops retook the fort, Lee was beaten. A few days later, Thomas Wesaw was struck in the face with a fragment from an exploding shell. A piece of iron struck below his left eye and pierced the bone. **He was one of the last battlefield casualties of Company K.** While Thomas Wesaw was in the hospital recovering, General Lee surrendered. The war was over!
Thomas Wesaw was discharged June 3, 1865 from the hospital at City Point, Va. S-pf-139
His wound healed and the scar was barely visible. But he gradually went blind in his left eye.
In 1884 he applied for a disability pension and received it. S-pf-139
There are at least three known photographs of Thomas Wesaw, taken after the war. S-ccn139a
TWO FAMILY STORIES: When the U.S. Army rounded up the Potawatomi in Michigan in 1840 and sent them west on the Trail Of Tears, Thomas Wesaw was a baby. He was accidentally left behind in a village. Other Indians who escaped the roundup heard him crying and rescued him. While Tom recovered from his wound, Abraham Lincoln visited the hospital and shook his hand. Both stories have some documentation, and in the opinion of the author, are credible. S-ccn139b

JOHN WESLEY 140

Wesler, or Wesley, John. Enlisted in company K, First Sharpshooters, Feb. 1, 1864, at Kalamazoo, for 3 years, age 28. Mustered Feb. 2, 1864. Promoted Corporal March 1, 1865. Mustered out at Delaney House, D. C., July 28, 1865.

Wesley, John – PRIME NAME, from Pension File
Wessley, John – State Death Record name
Naw-yaw-waw-sung – S-1870 Indian Annuity & Durant Roll No. 7348, Field Notes #27-pg52

RESIDENCES: Born on the Grand river where Grand Rapids is today S-pf-140
From 1865-1908: First he lived in Elbridge township, Oceana County, then he moved to Grant township, Mason County S-pf-140
DEATH DATE: September 26, 1908 (see note below) S-State Death Record
GRAVE: Mason County, Grant township / Grant Township Cemetery (see note below) *His unmarked grave is located either in Grant Cemetery on Free Soil road (aka Hansen road, aka Decker road), between US-31 and LaSalle Road, or Maple Grove cemetery on Free Soil Road between Darr road and Tuttle road.*
PENSION FILE: Yes, see Appendix A; 1883 Pension Roll: No
VETERAN CENSUS: 1888 No / 1890 No / 1894 Yes
RAY HEREK's BOOK: Pages: 435(Roster)

SPECIAL INFORMATION: He was born on the Grand River in Grandville (S-pf-140), which today is part of Grand Rapids. This indicates he was a member of one of the Grand River bands. He enlisted in February 1864 at Schoolcraft in Kalamazoo County. S-Descriptive Roll, State Archives. When the Civil War ended, John Wesley settled in Elbridge township, Oceana County. He married Mary Bailey there in 1866. Three children lived to be adults S-pf-140. Some years later (before 1889) he moved to Mason County and settled in Grant township, near the town of Free Soil. He died there in 1908 from Consumption (tuberculosis) S-pf-140 & State Death Record.
John's son, Frank Wesley (1884-1978) was a carpenter. When Kewadin, Antrim Co., celebrated its centennial (1965), Frank made wood crosses for the graves in Kewadin Indian Mission cemetery. Frank's grandson, Melvin Wanageshik, was killed in action in 1968 in the Vietnam war. His name is on The Wall monument in Washington, D.C. His name also appears on the Michigan Vietnam Memorial in Lansing. Melvin was the great-grandson of John Wesley.
NOTES FOR HISTORIANS: The Adjutant General Record gives two versions of name, Wesler and Wesley. **Since his pension file uses the name Wesley, that is his correct name.**
Note: Both the Mason County Death Record and State Death Record spell his name "Wessley". In his pension file is a certified copy of a Death Certificate written by the Mason County clerk, which says John Wesley died September 19, 1908. However, the State Death Record says he died September 26, 1908. In my judgment, the State Record is correct. His obituary confirms that he died Sept. 26. S-Ludington Chronicle, Sept. 30, 1908 John Wesley's obituary and his State Death Record say he is buried in Grant Cemetery. However, the obituary for his wife and other records indicate he is buried in nearby Maple Grove Cemetery. S-Ludington Daily News, July 12, 1925

CHARLES WHITEFACE 141

Whiteface, Charles, Pentwater. Enlisted in company K, First Sharpshooters, July 4, 1863, at Elbridge, for 3 years, age 22. Mustered July 11, 1863. Died of disease at Camp Douglas, Chicago, Ill., Feb. 14, 1864.

Whiteface, Charles – PRIME NAME, from Adjutant General Record
Whitbrain, C V – Gravestone name
Now-quo-um, Charles – Pension File, aka name
Wah-boose – Pension File, name used by his family and friends

RESIDENCE IN 1863: Pentwater, Oceana County S-AGR

DEATH DATE: February 14, 1864

GRAVE: Illinois, Cook Co., Chicago / Rosehill Cemetery – Section A (Military stone)
Buried in the Main Soldiers Lot, near the main entrance of cemetery, as "C. V. Whitbrain".

PENSION FILE: Yes, see Appendix A; 1883 Pension Roll: No

VETERAN CENSUS: 1888 No / 1890 No / 1894 No
RAY HEREK's BOOK: Pages: 419, 435(Roster)

SPECIAL INFORMATION:
Born near Ionia around 1841, Charley lived along the Grand River when it was still a wilderness. During his childhood his tribe, the Grand River Band, was left alone to live in their traditional way. But that ended when his chief signed a treaty. In 1858, at age 17, Charley and his tribe moved to a reservation near Hart, Michigan. His nickname was Wah-boose (Rabbit). His formal name was Now-quo-um. In the nearby town of Hart, folks called him Charley Whiteface. In time he fell in love with a girl named Susan. The two were married according to Indian custom. But things did not work out; she left him, taking their baby daughter. Charley was devastated by the failure of his marriage. When a recruiter for the U.S. Army visited the reservation a short time later, he enlisted. Saying goodbye to his father and sister, he went off to fight in the Civil War. He never returned. At his first duty station, Camp Douglas, Chicago, he contracted typhoid pneumonia from the prisoners he was guarding and died from the disease before seeing his first battle. On his gravestone the army wrote "C. V. Whitbrain". He is buried in Rosehill Cemetery. *Sources for the above story are the facts in his pension file, and various histories of Ionia County.* **Hart, Michigan, has a new Veterans Monument. If you have time, stop and see it. Among the names of local men who died in wars, you will see "Charles Now-quo-um Whiteface".**

NOTE TO HISTORIANS: The Adjutant General Record says he died February 14, 1864, and this is probably correct. But in his pension file is a document from the Army Surgeon General saying he entered the hospital on Feb. 14, and died the next day on February 15. Further research needed.

JAMES WILLIAMS 142

Williams, James, Lapeer. Enlisted in company K, First Sharpshooters, May 29, 1863, at Dearborn, for 3 years, age 18. Mustered June 16, 1863. Deserted July 5, 1863.

Williams, James – PRIME NAME, all sources agree on spelling

RESIDENCE IN 1863: Lapeer, Lapeer County S-Descriptive Roll, State Archives

DEATH DATE: Unknown

GRAVE: Location Unknown

PENSION FILE: No Pension File exists

VETERAN CENSUS: 1888 No / 1890 No / 1894 No

RAY HEREK's BOOK: Pages: 435(Roster)

SPECIAL INFORMATION:
James and Samuel Williams enlisted together into Company K on the same day.
It is likely they were brothers or cousins. They deserted together three weeks later.

 All Rights Reserved by Chris Czopek

MOSES WILLIAMS 143

Williams, Moses. Enlisted in company K, First Sharpshooters, July 4, 1863, at Marquette, for 3 years, age 30. Mustered July 20, 1863. Killed in action before Petersburg, Va., July 30, 1864.

Williams, Moses – PRIME NAME, from Pension File

RESIDENCE IN 1863: La Pointe, Ashland Co., Wisconsin (For source, see notes below)
(*He enlisted in Marquette, but his home was in Ashland Co., Wisconsin - see note below*)

DEATH DATE: July 30, 1864 (*KIA at the Battle of the Crater, Petersburg, VA*)
Note: KIA = Killed in action.

GRAVE: Location Unknown

PENSION FILE: No Pension File exists

VETERAN CENSUS: 1888 No / 1890 No / 1894 No

RAY HEREK's BOOK: Pages: 234, 351, 402, 435(Roster)

SPECIAL INFORMATION:
This soldier joined the army on the 4th of July.
He enlisted in Marquette, a town in Michigan's Upper Peninsula.
Army records say his home was in Ashland County, Wisconsin. S-Adjutant General Annual Report
Moses Williams was killed in action at the Battle of the Crater on July 30, 1864.
This was the most famous event of the Siege of Petersburg. The story begins with a regiment of Pennsylvania soldiers coming up with a daring plan. The men had been coal miners before the war. They figured they could dig a tunnel under the enemy line at Petersburg, put barrels of gunpowder at the end of the tunnel, and blow up the fortifications. It seemed like a good idea, so they received permission to try.
The fuse was lit July 30, 1864. The explosion made a huge crater!
And after that, everything went wrong.
General Grant later called the battle the saddest thing he saw in the whole, entire war.
Moses Williams was killed in the fighting at "the Crater", Petersburg, VA.

NOTE TO HISTORIANS:
Since the only other Company K soldier from Ashland County, Neveaux, was from La Pointe, and since Neveaux was mustered into Company K on the same day as Williams, it is likely that Moses Williams was also from La Pointe. Further research needed to verify this assumption.

SAMUEL WILLIAMS 144

Williams, Samuel, Lapeer. Enlisted in company K, First Sharpshooters, May 29, 1863, at Dearborn, for 3 years, age 20. Mustered June 16, 1863. Deserted July 5, 1863.

Williams, Samuel – PRIME NAME, all sources agree on spelling

RESIDENCE IN 1863: Lapeer, Lapeer County S-Descriptive Roll, State Archives

DEATH DATE: Unknown

GRAVE: Location Unknown

PENSION FILE: No Pension File exists

VETERAN CENSUS: 1888 No / 1890 No / 1894 No

RAY HEREK's BOOK: Pages: 435(Roster)

SPECIAL INFORMATION:
James and Samuel Williams enlisted together into Company K on the same day.
It is likely they were brothers or cousins. They both deserted three weeks later, on the same day. Nothing else is known about them.

 All Rights Reserved by Chris Czopek

PAYSON WOLF 145

Wolfe, Paysow, Northport. Enlisted in company K, First Sharpshooters, Aug. 3, 1863, at Northport, for 3 years, age 30. Mustered Aug. 8, 1863. Taken prisoner at Petersburg, Va., June 17, 1864. Discharged at Camp Chase, Ohio., June 13, 1865.

Wolf, Payson – PRIME NAME, from Pension File
Wolfe, Paysow – Adjutant General Record name (see note below)

RESIDENCE IN 1863: Northport, Leelanau County S-AGR
In 1883: Cross Village, Emmet County S-1883 Pension Roll
DEATH DATE: December 7, 1900 S-State Death Record
GRAVE: Emmet County, Cross Village, Holy Cross Cemetery (Military stone)
(*Location: Holy Cross Cemetery is on Levering road, exactly one mile east of Cross Village*)
PENSION FILE: Yes, see Appendix A
1883 Pension Roll: Yes, Emmet Co. - Cross Village (Loss use left hand and arm)
VETERAN CENSUS: 1888 No / 1890 Yes / 1894 Yes
RAY HEREK's BOOK: Pages: 34, 288, 290, 372, 399, 435(Roster), 495
SPECIAL INFORMATION:
He was a full blood Odawa (Ottawa) who grew up near Traverse City on the Leelanau peninsula. Chief Peter Waukazoo was his uncle. Payson was given an education by the Rev. George Smith, a missionary living near Northport. Eventually he married one of the missionary's daughters. The Rev. Smith kept a diary, which is a source of useful information about his remarkable son-in-law. During the Civil War, Payson left his farm and family to enlist in Company K. He was proud to be a soldier – **the only picture in existence taken during the Civil War of a Company K soldier in uniform, is a picture of Pvt. Payson Wolf** (S-ccn145a). He named one of his sons after Captain Edwin Andress, the popular commander of Company K.
Payson Wolf was captured and sent to Andersonville prison. He managed to survive and return home. But his time in the Confederate prison took a toll on his health. He was unable to use his left hand and arm for work on the farm, so he asked the government for a disability pension.
Payson was a member of the Washington G.A.R. Post No. 106, Cross Village, Emmet County. He appears in a group picture of the Woolsey G.A.R. Post No. 399, Northport, Leelanau Co. The GAR Medal he wears in the picture is now on display in the Northport history museum.
NOTES FOR HISTORIANS:
Payson's name cannot be found on any of the membership lists for Northport GAR Post No. 399. Correct spelling of his first name is Payson. It means "I leave this thing behind" (S-ccn145b). His last name is Wolf, because his father's last name was Mi-in-gun, which means Wolf. Payson signed his name "Wolf" without an "e" at the end, so that is his PRIME NAME. The Adjutant General Record spells it Wolfe, with an "e" at the end, as do all his descendants living today.
See page 69 of the book Rethinking Michigan Indian History, published 2005 by Patrick LeBeau. Dr. Susan E. Gray of the History Department of Arizona State University has written a scholarly study of the life of Payson Wolf, titled Miengun's Children, published 2008 in FRONTIERS. See the manuscript at The Holland Museum Archives and Research Library in Holland, Michigan, written by Payson's daughter, Etta S. Wilson, about Payson's Odawa parents. Great reading!

JOHN KEDGNAL 146

Kedgnot, or Kegnot, John. Enlisted in company B, First Sharpshooters, Dec. 10, 1862, at Grand Haven, for 3 years, age 18. Mustered March 31, 1863. Taken prisoner at Reams' Station, Va., Aug. 26, 1864. Discharged at Detroit, Mich., May 31, 1865. Present residence, Petoskey, Mich.

Kedgnal, John – PRIME NAME, from Pension File
Kedgnot, John – Adjutant General Record name
Kegnot, John – Adjutant General Record, aka name
Duvernay, John Henry – Gravestone name
Deverney, John – 1894 Veteran Census name

RESIDENCE IN 1862: Grand Haven, Ottawa County S-AGR
In 1894: Petoskey, Emmet County S-1894 Veteran Census
DEATH DATE: April 2, 1926 S-Greenwood Cemetery records
GRAVE: Emmet Co., Petoskey / Greenwood Cemetery - Section H, Block 36, Grave 3
(His gravestone is a modern bronze plaque that lies flat on the ground)

PENSION FILE: Yes, see Appendix A; 1883 Pension Roll: No

VETERAN CENSUS: 1888 No / 1890 No / 1894 Yes
RAY HEREK's BOOK: Pages 21, 246, 404, 423(Roster of Company B), 489

SPECIAL INFORMATION:
He belonged to the Chippewa tribe, according his statement in the Albert Pesherbay pension file.
His sister, Louisa, married Company K soldier Louis Genereau Jr. S-pf-036 (Genereau pension file)
John Kedgnal, aka John Henry Duvernay, served in Company B. Yes, he was a Native American in the 1st Michigan Sharpshooters, but he was NOT in Company K.
John Kedgnal suffered much in the war. He almost died from Typhoid fever. He was wounded in the right leg near the knee. And he was a POW at Salisbury Prison in North Carolina. This prison was not as bad as the infamous Andersonville prison, but it was not much better. When released, John weighed only 80 pounds. A prison guard struck him in the back with a rifle butt, causing an injury that bothered him for the rest of his life. S-pf-146
John Kedgnal changed his name to **John Henry Duvernay** after the war. S-name on his gravestone
His half-brother, William Duvernay, enlisted in Company B as a drummer boy. S-Herek pg 21
These two brothers were the only Native Americans in the 1st Michigan Sharpshooters who were not members of Company K.
John Kedgnal's original discharge paper exists. It was passed down in his family. S-ccn146a
John was a member of the G.A.R., Lombard Post No. 170, Petoskey, Emmet County.
He appears in a group picture of this G.A.R. post, taken in 1921.

WILLIAM DUVERNAY 147

Duverney, William. Enlisted in company B, First Sharpshooters, as Musician, Dec. 4, 1862, at Grand Haven, for 3 years, age 13. Mustered March 31, 1863. Mustered out at Delaney House, D. C., July 28, 1865.

Duvernay, William – PRIME NAME, from Pension File
Duverney, William – Adjutant General Record name
"Chid" Duvernay - His nickname in Grand Haven, after the war.

RESIDENCE IN 1862: Grand Haven, Ottawa County S-AGR
In 1890: National Soldiers Home in Milwaukee, Wisconsin S-1890 Veteran Census

DEATH DATE: August 25, 1893 S-pf-147 *(died at the Soldiers Home in Milwaukee)*
GRAVE: Ottawa Co., Grand Haven / Lake Forest Cemetery, Section 7 (Military stone)

PENSION FILE: Yes, see Appendix A; 1883 Pension Roll: No

VETERAN CENSUS: 1888 No / 1890 Yes / 1894 No

RAY HEREK's BOOK: Pages: 21, 237, 246, 250, 368, 423(Roster of Co.B)

SPECIAL INFORMATION:
William Duvernay was a Native American living in Grand Haven, Michigan. He enlisted in Company B as a drummer boy at the age of 13.
John Kedgnal of Grand Haven enlisted about the same time he did. They were half-brothers. Years later, John changed his last name to Duvernay (S-Herek pg 21, and name on John's gravestone).
Company B took a liking to the little Indian drummer boy and his brother John. In fact, they were so popular, that when Company K was being recruited, the Colonel wanted to transfer them, but the men in Company B refused to give them up! They told the Commander, "These are OUR Indians, and we're not letting them go!". **So that is why these two brothers were the only Native Americans in the 1st Michigan Sharpshooters who were not members of Company K.** S-ccn147a
William Duvernay never fired a rifle in battle. But he saw all the horrors of war, first as a drummer boy, and later as a Hospital Attendant. When the war ended, he returned home to Grand Haven.
As a talented musician who could play many instruments, he was a well-known and well liked member of the local bands. His nickname was "Chid". A picture exists of him holding a violin and wearing his Civil War hat (S-ccn147b). His life was shortened by a disease (dysentery) from his army days. While being cared for at the National Soldiers Home in Milwaukee, Wisconsin, friends rode the ferry boat across Lake Michigan to visit him. Reports of these visits were printed in the Grand Haven newspaper. The last report said, his dying wish was to be buried in the Duvernay family plot in Grand Haven (S-ccn147c).
According to his obituary, William had lied to the recruiter about his age. He was actually 11 years old when he enlisted. He may be **the youngest soldier in the Union army during the Civil War!**

FOOTNOTES

My footnotes in the Facts and Fate section of this book are written as S-ccn. The S- means "Source", ccn means "Chris Collection Note", indicating the information comes from the author's notes.

ABREVIATIONS for SOURCES

S- = Source
S-AGR = Source: Adjutant General Record
S-pf-131 = Source: Pension File for soldier No. 131
(Soldier Numbers are the CL- numbers in this book's Roster of Company K section.)
S-Herek pg 388 = Source: Raymond Herek's book "These Men Have Seen Hard Service" 1998, Wayne State Press, pg. 388
S-ccn001A = Source: Chris Collection Note, number 001a.

FOOTNOTES :

ccn001a Charles J. Meyers, well known Michigan historian who did valuable research on Company K, says the name Agahgo means "Old Porcupine".

ccn002a Magazine: Civil War Times, June 2007, "Inured to Hardships, Fleet as Deer" by Gordon Berg. On pg. 55 is an excellent article telling the story of Company K; it mentions Charles Allen.

ccn002b Missionary George N. Smith lived near Northport on the Leelanau peninsula. His daily diary mentions George Allen and other men who joined Company K. The diary has never been published, but a transcript was loaned to me by Avis Wolfe of Northport, Michigan.

ccn004a A descendant of Peter Anderson told me, Peter was probably buried at the Kewadin Indian Mission cemetery.

ccn004b The information about his family and his band was provided by a descendant of Peter Anderson.

ccn005a In Wildwood Cemetery, Chesaning, Michigan, the Andress family plot has an obelisk with the following inscription: Edwin V. Andress, born July 29, 1838. Died in Weldon, Cal. Oct. 10, 1884.

ccn005b Credit for the discovery of Captain Andress in the 1864 Marye's Heights photograph goes to Lansing historian Roger Winthrop.

ccn007a After the Civil War, the Quartermaster General of the army published a master list of all known graves of Union soldiers. It was called THE ROLL OF HONOR. The name "O. Ahreotergeshig" appears in Vol. 1, pg 107, on the list of graves in Arlington Cemetery. It is a misspelling of the name Oliver Aptargeshick.

 All Rights Reserved by Chris Czopek

ccn008a Traverse City Record Eagle, April 20, 1905.

ccn008b This information on residences of James Arwonogezice is from records of the "Old Soldiers Home" in Grand Rapids. He appears in the "Soldiers Home Record Book" as resident # 3742. A copy of this book can be found at State Library & Archives, Lansing.

ccn008c There is a booklet in the collection of the State of Michigan Library in Lansing (call number E 514.5, 1st , F57 .Z) printed for the 1886 Reunion of the 1st Michigan Sharpshooters. On pg. 15 is a list of the members of Company K who attended the reunion.

ccn008d Conflicting information on grave location: Charlevoix Sentinel May 24, 1917 front page, says he was buried in an Indian Cemetery near Bay Shore. Charlevoix Courier, May 23, 1917 pg5 says he was buried at Susan Lake, Charlevoix [= Greensky Hill Indian Mission burial ground]. However, his State Death Record says his grave is in Brookside Cemetery, City of Charlevoix.

ccn010a In the collection of the author is information copied from an 1872 land patent.

ccn010b In November 2005, I visited Onominise cemetery. It was overgrown and neglected. The gravestone for George Ashkebug was out of the ground and leaning against the side of a small hill at the edge of the cemetery. The Iron GAR marker stood alone in the middle of the cemetery. It was my impression that the gravestone originally stood beside the GAR marker.

ccn011a His name appears as "Amos Green" on the list of survivors of the Sultana tragedy, published on front page of Detroit Advertiser & Tribune May 24, 1865. He is identified as belonging to the 1st Lt Art Co.K, which is a mistake.

ccn011b Records of the Scottville GAR Post No. 348 can be found in the Michigan State Archives, Lansing MI.

ccn013a After the Civil War, the Quartermaster General of the army published a master list of all known graves of Union soldiers. It was called THE ROLL OF HONOR. This soldier appears in Vol. 13, pg 67. A note there says he was originally buried at the cemetery of the Ninth Corps [hospital] at Meade Station.

ccn014a **Adjutant General Annual Report** - Once a year, every year, the Adjutant General of the State of Michigan issues an Annual Report. During the Civil War years, these Annual Reports listed every Michigan soldier who died, or was wounded, captured, or discharged. Some of the information differs from the Adjutant General Record, which was published around 1907, more than forty years after the end of the war. The Annual Reports are a valuable primary source known to only a few Civil War historians. They have never been reprinted. The only copies in existence are at State Library in Lansing, call number MDOC UA 43 .M5.

ccn029a Letter from historian Raymond Herek says Amos Crane was buried in the Pinery. He was given the information by an amateur historian. Besides that, there is really no other place an Indian who died in Zeba could be buried.

ccn032a The soldiers of Company K felt honored to have the brother of the regiment Commander as their Captain. This was told to me by Charles J. Meyers, expert on the history of Company K.

ccn036a According to an old map given to me by Frank Kequom of Mt. Pleasant. Frank was born and raised in Elbridge township, Oceana County.

ccn039a The Fredericksburg National Cemetery website says: (Grave 2186) Gora, Samuel Private Co.K, 1 MI. Died 20 May, 1864. Originally buried at Marye's Heights, Fredericksburg, Virginia.

ccn040a Obituary in a Detroit newspaper, Advertiser and Tribune, date unknown. My information comes from a copy of this obituary in the pension file of Garrett Graveraet.

 All Rights Reserved by Chris Czopek

ccn040b The old <u>Graveraet High School</u> in Marquette, built around 1926, was named after Juliet Graveraet, who was the aunt of Lt. Garrett Graveraet, and also the sister of Company K soldier Henry G. Graveraet Jr. Source: Charles J. Meyers.

ccn041a "Garrett A. Graveraet was the son of Bear River band chief Mankewenan, Henry G. Graveraet, a Franco-Ottawa merchant-trader at Mackinac Island …" This is a quotation from a book published in 1995, "Between Two Fires" by Laurence Hauptman.

ccn041b The old <u>Graveraet High School</u> in Marquette, built around 1926, was named after Juliet Graveraet, who was the sister of Company K soldier Henry G. Graveraet Jr., and who was also the Aunt of Lt. Garrett Graveraet. Source: Charles J. Meyers.

ccn044a Exact source of this information, and his family tree, is available from the author.

ccn045a Source of this information is available from the author.

ccn046a The 1868 Annuity Roll of Saginaw and Chippewa (see Lantz, "Ottawa and Chippewa Indians of Michigan 1855 - 1868", pg. 93) shows his family were members of Mackinac Band.
The 1850 U.S. Census shows that Michael Jeandron was the brother of Rosalie, and Rosalie was the wife of James Hamlin, making James and Michael brothers-in-law.

ccn050a He had a wife in Bay Shore at the time of his death, so apparently his home was there. See the newspaper story about his death in the Boyne Citizen, Tuesday, March 26, 1907, front page.

ccn050b Location of this army discharge paper is available on request.

ccn051a State Death Record says he died April 21, 1907 in Evangeline township, Charlevoix County.

ccn051b In John Jacko's pension file is an official copy of his death certificate, made by the Charlevoix County clerk in 1911, that says he died April 4, 1907. However, the State Death Record card for John Jacko says he died April 21, 1907, and this date is confirmed by newspaper stories about his death. So the correct date of death is April 21, 1907.

ccn051c John Jacko had a brother named Louis who died in Detroit in 1870 and was buried in Elmwood Cemetery under the name Jacko Nat-bah-me-no-ling. When the cemetery ordered a new gravestone, they could not find his name on Adjutant General Record, so they assumed he was John Jacko and ordered a stone with that name. This was a mistake; John Jocko is buried in Boyne City. Documents proving Louis Penaiswanquot is Jacko Nattahmenoting can be found in the pension file for Company K soldier Jacko Penaiswonquot.

ccn053a This information comes from correspondence with a descendant. Name and address available on request.

ccn053b Source of this information is a copy of his death certificate, which I found in his pension file.

ccn055a **The Story of the Atwater List** : A prisoner at Andersonville named Dorence Atwater was selected to keep a record of prisoners who died. He made a secret copy of the list and carried it with him when he was released at the end of the war. It was published in 1866 as a book. This is what I call "The Atwater List". It is the only true source of information on the dates when soldiers died at Andersonville. The Library of the State of Michigan has a copy of this rare book. Look for the title "A List Of The Union Soldiers Buried At Andersonville", Dorence Atwater, 1866.

ccn055b The 1868 Annuity Roll of Saginaw and Chippewa (Lantz, Ottawa and Chippewa Indians of Michigan 1855-1868, pg. 93) shows his family were members of Mackinac Band. The 1850 U.S. Census shows that Michael Jeandron was the brother of Rosalie, the wife of James Hamlin, making James and Michael brothers-in-law.

ccn055c "Before the Bridge", a book on St. Ignace history published by the Kiwanis Club in 1957. On pg 54 it names the earliest settlers St Ignace. One name is Michael Jaudron Sr. The book says his aka names are Jeandron and Jeandreau. This book also says, that Michael Jeandron Sr. had a daughter Marianna who married a Perrault. The 1850 U.S. Census shows that Perrault and Maryanne were parents of Michael Jeandron, the soldier in Company K.

ccn058a Date of death and cause of death come from the "Oceana County Native American Death Index 1867-1875".

ccn059a This spelling of his name is from the "History of Saginaw County", a book published in 1881 by the Chapman Publishing Co., Chicago.

ccn061a The book "Michigan Place Names" by Walter Romig says Clare County was first called Kaykakee County after an Indian chief by that name, and the name means Pigeon Hawk.

ccn061b On his Death Certificate, his name is Joseph Me-she-ka-kuch. On the 1894 Veteran Census, his name is Joseph Me-she-ka-kack.

ccn063a There is a booklet at State of Michigan Library in Lansing (call number E 514.5, 1st , F57 .Z) printed for the 1886 Reunion of the 1st Michigan Sharpshooters. On pg. 15 is a list of the members of Company K who attended the reunion. His name is on the list.

ccn063b This information is from Grayling's newspaper, the Crawford County Avalanche, Aug 21, 2003 - The Historical Keepsake Edition. There is a copy at the Clarke Library in Mt. Pleasant in the file folder for Thomas Kechittigo.

ccn063c A copy of this photo is in the collection of the author. Original photo is in a collection outside of Michigan.

ccn065a This information comes from an expert genealogist, name available on request.

ccn067a Two facts indicate Benjamin Kewaconda belonged to Bear Creek Band: The 1860 census shows three families named Kewacondo living in Bear Creek township, Emmet county, and the 1855 Ottawa and Chippewa Annuity Roll shows payouts to members of the Bear Creek Band named Ke-way-cun-do. But it is not certain Benjamin was related – further research needed.

ccn068a According to his pension file, there was an inquiry in 1899, and the government concluded that March 3 was his correct death date. The decision was based on medical records.

ccn071a For details of this story, contact the author.

ccn078a Diary of Rev. George N. Smith, missionary at Northport, Michigan. Original diary is in the collection of the Library of Congress, Washington, D.C.

ccn078b Letter written by Payson Wolf to his wife, dated Jan 16, 1864. He was stationed at Camp Douglas and says "Osah-o-quah-yon & Kewagoshkom are getting well". My conclusion is, these two are Adam Sawbecome and Louis Miskoguon. The original letter was in the collection of Bud and Avis Wolfe of Northport, Michigan. Present location of letter is unknown.

ccn078c The Detroit Advertiser & Tribune May 24, 1865, published a list of names of survivors of the Sultana steamboat explosion on the front page. Louis Miskoguon appears as "L. Scogen".

ccn078d Contact the author for the source of this information.

ccn079a Diary of Rev. George N. Smith, missionary at Northport, Michigan. Original diary is in the collection of the Library of Congress, Washington, D.C.

ccn079b **The Story of the Atwater List** : A prisoner at Andersonville named Dorence Atwater was selected to keep a record of prisoners who died. He made a secret copy of the list and carried it with him when he was released at the end of the war. It was published in 1866 as a book. This is what I call "The Atwater List". It is the only true source of information on the dates when soldiers died at Andersonville. The Library of the State of Michigan has a copy of this book. Look for the title "A List Of The Union Soldiers Buried At Andersonville", Dorence Atwater, 1866.

ccn082a Book "These Men Have Seen Hard Service; The First Michigan Sharpshooters in the Civil War" by Raymond J. Herek, 1998, Wayne State University Press.

ccn087a After the war, Joseph Nesogot must have lived and died in Taymouth township, Saginaw County, because his grave is there.

ccn087b It was common for an Anishinabe son to use his father's first name as a last name. Example: John, the son of Jacko Penaiswonquot called himself "John Jacko". This could explain how Joseph Nesogot became Joseph William(s).

ccn088a According to Lansing historian Roger Winthrop, the grave of William Neveaux is at La Pointe, Wisconsin (Ashland County, on Madeline Island) and in the 1960s a military stone was placed.

ccn088b Source of information about William Neveaux on the Color Guard, his serious illness, and his temporary service as a hospital attendant is found in the Descriptive Roll, State Archives.

ccn088c Lansing historian Roger Winthrop also discovered Company K soldier Mark Peshekee is buried at La Pointe, Wisconsin, in the same cemetery as William Neveaux.

ccn090a There is a booklet at the State of Michigan Library in Lansing, call number E 514.5, 1st , F57 .Z printed for the 1886 Reunion of the 1st Michigan Sharpshooters. On page 15 is a list of the members of Company K who attended the reunion.

ccn090b The author has carefully studied the family tree of Asahel Nichols, and proved there is no relation between Asahel Nichols and Lemuel R. Nichols of the 1st Michigan Sharpshooters.

ccn091a Kirke is correct spelling of his first name. However, his obituary and his State Death Record both spell his first name "Kirk" without an "e".

ccn094a The grandsons of Marcus Otto have told the author that Marcus is buried on the family farm near Weidman (Isabella County), and that they know the exact place.

ccn095a The pension file of Marcus Otto says he was paid a $25 bounty. Since Marcus and Solomon enlisted same place, same day, they both must have received the same bounty.

ccn095b The story of Solomon Otto being sent home to Michigan was told to me by the grandsons of Marcus Otto. Their father told them that Solomon died either on the way back to Michigan, or in a hospital in Detroit.

ccn095c The author has a copy of the 1871 Land Allotment certificate, made out to Solomon Otto.

ccn098a Death date comes from the Atwater List. The Story of the Atwater List : A prisoner at Andersonville named Dorence Atwater was selected to keep a record of prisoners who died. He made a secret copy of the list and carried it with him when he was released at the end of the war. It was published in 1866 as a book. This is what I call "The Atwater List". It is the only true source of information on the dates when soldiers died at Andersonville. The Library of the State of Michigan has a copy of this rare book. To see it, look for the title "A List Of The Union Soldiers Buried At Andersonville", Dorence Atwater, 1866.

ccn099a According to Lansing historian Roger Winthrop, the grave of Mark Peshekee is at La Pointe, Wisconsin (Ashland County, on Madeline Island).

ccn100a This information comes from correspondence with Art Dembinski M.D. (retired), a descendant of Company K soldier Albert Pesherbay.

ccn103a He was discharged before his enlistment ran out, a sure sign he was a disabled soldier.
The 1883 Pension Roll says he was wounded in the left hip.
He collected his pension from 1866 to 1906 – total of forty years!

ccn104a Source of information: "Indians of Ontonagon County In Company K", an article printed in the Kalamazoo Family Newsletter, Vol. 7, 1977-1978, pg 110.
About John Ruberdee it says "On discharge he retained his ordinance (rifle) for a payment of 6.00." Looks like he took his army rifle home.

ccn105a The Grand Rapids Public Museum has a photograph on display of "Aaron Pequongway" and his family. The picture was taken in 1915. It is part of a Native American Veterans exhibit.

ccn105b The newspaper Traverse City Record Eagle, October 29, 1955, has an interview with Omena resident W.A. Craker, an old pioneer. He says that Aaron was with General Sherman on the famous March to the Sea.

ccn107a "History of Saginaw County", a book published 1881 by the Chapman Publishing Co., Chicago.

ccn107b George W. Lee was the Indian Agent for Michigan. He helped the widow of Sashkobanquot apply for a pension. For details, see the 1879 – 1880 correspondence between Indian Agent George W. Lee and the Commissioner of Indian Affairs in Washington D.C.

ccn107c Source of this particular spelling of his name available from author on request.

ccn108a Letter written by Payson Wolf to his wife, dated Jan 16, 1864. He was stationed at Camp Douglas and says "Osah-o-quah-yon & Kewagoshkom are getting well". In my opinion, these two men are Adam Sawbecome and Louis Miskoguon. The original letter was in the collection of Bud and Avis Wolfe of Northport, Michigan. Present location of the original letter is unknown.

ccn109a In 1887, the Adjutant General Office of the War Department attached a note to the Military Record of Antoine Scott, showing that Major General John G. Parke, commander of the 9th Army Corps, had recommended Antoine Scott for the Medal of Honor on February 21, 1865. The note is "Bookmark 1339-A-1887". Also, since Antoine Scott died nine year earlier in 1878, it is possible he was unaware of the honor he had earned.

ccn116a When he was wounded in action June 17, 1864, he would have been immediately sent to a hospital. He was discharged a year later in June 5, 1865 from a hospital in Detroit. I feel safe assuming he was in hospitals all that time. Notice his discharge paper specifically says he was unfit for service in the V.R.C. [Veteran Reserve Corps]. He appears to have been badly wounded and had a long, difficult recovery.

ccn117a The two cousins were both named "John Baptiste Shomin". This information comes from the genealogy files of a respected genealogist who shared the information with me.

ccn118a The fact the two cousins were both named "John Baptiste Shomin" comes from the genealogy files of a respected genealogist who shared the information with me.

ccn123a He died December 13, 1890, in an accident. He went to sleep on the railroad tracks near his house while intoxicated and was run over by a train. This according to a document in State Archives, Lansing. Look for "Joint Documents of the State of Michigan" for the year 1890, Volume 4, Part 1, Nineteenth Annual Report of the Commissioner of Railroads, pg 245. On this page is a list of deaths in railroad accidents, and Freeman Sutton is mentioned.

 All Rights Reserved by Chris Czopek

ccn123b Since he applied for a pension in 1865 as soon as he was discharged, and the pension was granted in 1882, his leg must have been seriously injured and remained that way for the rest of his life.

ccn124a The name of the website displaying the picture of Francis Tabasash is available from the author on request.

ccn126a This information about Henry J. Pawneshing comes from a newspaper clipping in the collection of the Oceana County Historical Society, Hart, Michigan.

ccn126b The image on display at the Grand Rapids museum comes from a photo postcard in the Local History collection of the Grand Rapids Public Library, collection number "Coll. 068 (068-1-15)". There is a note on the back of the post card which says "Po-ne'-sing, prominent Ottawa Indian of Elbridge, Oceana Co., Mich."

ccn127a This information comes from descendants. Contact the author for their names.

ccn130a Grave Location: In his pension file, the widow of Charles Wabesis says she and Charles were married by a Catholic priest. This means they were Catholics. Since they lived in Elbridge township, I say he was probably buried in Elbridge township in St. Joseph Catholic cemetery.

ccn132a Art Dembinski M.D. (retired) is a descendant of Chief Waukazoo and has carefully studied the family genealogy. He asserts that after the war, Company K veteran Joseph Wakazoo spent the rest of his life (1865-1910) in Minnesota.

ccn132b Place of his death is from a church magazine "The Living Church Annual 1911". On page 145 it reports the death of the Rev. Joseph Wakazoo on the White Earth Reservation in Minnesota.

ccn132c Art Dembinski M.D. (retired) has done extensive research on the life of Joseph Wakazoo. He discovered in the Manuscript Collection of the Western Reserve Historical Society, Ohio, the records of the Twinsburg Institute which show Joseph Wakazoo was a student there.

ccn132d The Descriptive Roll in State Archives, Lansing, specifically says Joseph Wakazoo was transferred from the 16th Michigan Infantry to Company K, 1st Michigan Sharpshooters, "by Order of the War Department".

Ccn132e A portrait of the Rev. Joseph Waukazoo, dated 1881, appears in The Gospel in All Lands: An Illustrated Monthly Missionary Journal, Vol XI, 1885, p. 312, published by the Methodist Episcopal Church Missionary Society.

ccn133a See Chapter 3 titled "The Great Copper Boulder from Ontonagon", in the book "This Land, The Ontonagon", by Bruce H. Johanson. 1984. Iron Mountain, MI. Small Press publication.

ccn133b The story of Henry's family was written down as part of "The Indian Project" of Sister Macaria Murphy, a nun who taught school on the Bad River Reservation. She interviewed Henry's brother, John Condeacon, in 1943. To find a copy of this story, search for "John Condeacon, Chippewa Indian Historical Project, Archives of the Fransciscan Sisters of Perpetual Adoration, La Crosse, Wisconsin". Available on microfilm at the State Historical Society of Wisconsin's Area Research Center in Ashland, Wisconsin.

ccn133c For an example of Henry Condecon being appointed to a tribal committee, see the proceedings of the Bad River tribal council meeting held at the council lodge in Odanah, Wisconsin, on Thursday, September 16, 1909. You can read the details on pg. 538 of a report titled "Condition of Indian Affairs in Wisconsin", by United States Senate Committee on Indian Affairs, published in 1910.

ccn133d There is a picture of Henry Condeacon wearing his GAR badge on the cover of the tribal newsletter for the Bad River Band in Wisconsin. Name of the newsletter is "News From The Sloughs", and the date is November 1995 (Vol. 1, issue 7).

ccn134a The obituary of Amos Watson, son of James V. Watson, is interesting. You can find it in the Battle Creek Enquirer Oct 29, 1926, under the heading "DEATH CLAIMS INDIAN SCRIBE". It mentions that for years Amos wrote articles for the Athens Times newspaper under the name "Wah-nuh-Wah". Proof that he is the son of Company K soldier James Watson is found in the pension file of George Mogage. In it, Amos gives testimony that he visited his father in an Army Hospital when the Sharpshooters were in Chicago. Since Amos was born around 1848, he would have been around 16 years old when he made the trip in 1864.

ccn135a His descendants spelled the name Wah-Bah-Noo. Source, descendant Denny Denman, in conversation with the author.

ccn135b The 1890 Veteran Census has a note at the bottom of the page, saying they tried to contact "John Waw-Be-Naw" but found him gone. He had left the reservation to sell his baskets.

ccn137a According to the probate papers for the estate of Martha W. Collins, William was struck by a train somewhere between St. Charles and Saginaw, and died. This was in April 1897. A newspaper story about this accident exists, see the Chesaning Argus, April 3, 1897.

ccn137b His name appears on a list of Civil War soldiers buried in Riverside Cemetery, St. Charles. See the Civil War Centennial List of Civil War soldier graves, it is on microfilm at the State Archives in Lansing (See Archives, Centennial Records, Box 563).

ccn139a For information on existing photographs of Thomas Wesaw, contact the author.

ccn139b An old Potawatomi woman, Mrs. Sarah Isaac, told the story about "the lost papoose", and it was published in 1934. Source of story of Lincoln meeting Thomas Wesaw is a descendant, Anthony Foerster, who heard it from his Aunt Alberta. Contact the author for complete details.

ccn145a The photograph was taken in 1863, when Payson Wolf was stationed at Camp Douglas, Chicago. He is wearing a "Hardee" hat (Model 1858 Dress Hat) which was the standard army hat before the Civil War "Kepi" became popular. The feather in his hat is his own idea – the picture was made for the folks back home, and he wanted to show pride in his heritage. The original picture is in the Burton Historical Collection, Detroit, item number CSI-182

ccn145b According to the late Charles J. Meyers, a well-known Michigan historian who did valuable research on Company K, the name "Payson" means "I leave this thing behind". See pg. 69 of a book titled "Rethinking Michigan Indian History", 2005, by Patrick LeBeau.

ccn146a For a copy of the discharge paper of John Kedgnal, aka John Henry Duvernay, contact the author. He can put you in touch with the family.

ccn147a The story of the Duvernay brothers in Company B, and how popular they were with the men of that Company, was told to me by the late Charles J. Meyers, a well-known Michigan historian who did valuable research on Company K.

ccn147b The photograph of William Duvernay holding a violin and wearing a Civil War hat is in the collection of the Clarke Library, Mt. Pleasant, Michigan.

ccn147c The death of William Duvernay was front page news - see the Grand Haven newspaper, Evening Tribune, August 24 and August 25, 1893. These stories tell of his wish to be buried with his parents, and how he was only 11 years old when he enlisted in the army.

APPENDIX A

LIST OF PENSION FILES FOR SOLDIERS IN COMPANY K

When a soldier died in the war, his widow or parents could apply for a pension. The application papers contain valuable family information such as place of birth, name of wife, date and place of death. You can order copies of Civil War Pension Files from National Archives in Washington D.C.

The information below is copied from the Pension File Index Cards for Civil War Pensions. I used the microfilm copy of these cards at State Library and Archives, Lansing. Names, dates, and numbers have been carefully checked for accuracy. All names are spelled exactly as they appear on the index cards.

INVALID = Pension for the soldier, for reason of wounds or old age.
WIDOW / MOTHER / FATHER = Pension for Widow or Parents who depended on soldier for support.
The date, such as "1877 Jan 6", is the date the Application was received in Washington.
A-xxxxxx = Application number. C-xxxxx = Certificate number.
Use the Certificate Number to order a copy of a pension file from National Archives.
When you see "C (blank)" it means the space for the Certificate Number is blank. This indicates the person applied for a pension, but the claim was either dropped or rejected, therefore the person never collected a pension.
MINOR = a child under the age of 16.
Gdn = Guardian, person who has custody of the minor child.
The name of a state, Mich or Wisc or Mass or Minn or Canada, indicates the state the application was mailed from. This gives a clue where the soldier or widow was living at the time they applied for a pension.

001 **Agahgo, Charles**
WIDOW: Agahgo, Pahquah Equa [alias Lucy Pequechquay on 1890 Veteran Census] 1865 Nov 18 / A-115,521 C-156,567
MINOR: Agahgo, J. Tamekey or Tamahey (Gdn) 1870 Feb 4 / A-184,346 C-(blank)

002 **Allen, Charles**
MOTHER: Allen, Annie 1881 Mar 11 / A-280,616 C-251,466

003 **Amderling, Peter (alias) Anotagan, Peter**
MOTHER: Anotagan, Oshabens 1890 Dec 30 / A-493,206 C-(blank) / Wisc
FATHER: Anotagan, John B. 1894 Mar 16 / A-591,887 C-402,007 / Wisc

005 **Andress, Edwin V.**
INVALID: 1879 Sept 25 / A-311,699 C-201,283
WIDOW: Andress, Ellen 1892 Apr 20 / A-547,567 C-882,283 / Calif
MINOR: Andress, Ellen (Gdn) 1892 July 5 / A-554,030 C-379,908 / Calif

 All Rights Reserved by Chris Czopek

APPENDIX A – LIST OF PENSION FILES

006 **Andrew, John**
INVALID: 1882 Sept 26 / A-460,944 C-252,595 / Mich
WIDOW: Andrew, Elisa 1885 Mar 3 / A-324,169 C-(blank) / Mich
MINOR: Slater, Joseph D. (Gdn) 1887 Aug 16 / A-359,322 C-276,735 / Mich

008 **Arwonogezia, James**
INVALID: 1889 April 17 / A-699,457 C-527,059 / Mich

009 **Ashkanak, Joseph**
INVALID: 1877 Jan 6 / A-229,500 C-147,960
WIDOW: Ashkanak, Elizabeth 1914 June 16 / A-1,029,588 C-778,421 / Mich

010 **Ashkebug, George (alias) Estabrook, George**
INVALID: 1884 Mar 22 / A-508,907 C-883,425
WIDOW: Ash-ke-bug, Mary 1889 Apr 8 / A-392,567 C-395,613 / Mich

011 **Ashkebugnekay, Amos**
INVALID: 1883 Mar 6 / A-474,654 C-344,840 / Mich

013 **Awanakwad, Betois (alias) Awanakwad, Betoes (alias) Pe-to-zo-ourquitte**
WIDOW: Awanakwad, Teresa 1865 Oct 19 / A-113,206 C-128,324
MOTHER: Ahpetahgshegoquay, Mary A. 1900 Apr 26 / A-718,265 C-(blank) / Mich

016 **Bennett, Louis**
MINOR: Coles, Mary J. 1891 Sept 14 / A-529,495 C-(blank) / Mich

019 **Cay-bar-coung, William (alias) Cowing Cayba, William (alias) Cowing, William C.**
INVALID: 1890 Aug 28 / A-930,155 C-(blank) / Mich
MINOR: Carson, SS [Samuel S. Carson] (Gdn) 1893 Feb 28 / A-576,105 C-(blank)

021 **Chamberlain, Amos**
INVALID: 1882 Sept 26 / A-460,960 C-399,643 [or C-379,643] / Mich
WIDOW: Chamberlain, Jane 1909 Oct 26 / A-929,391 C-(blank) / Mich
Remarks: X.C.926,542 **Note on card says: Died Aug 18, 1909.**

022 **Chatfield, Charles**
FATHER: Chatfield, Thomas 1892 Feb 23 / A-548,851 C-(blank)
Pension file says Mich 10 INF, Co.B. Same man as was in Mich 1 SS Co. K ?

023 **Chatfield, Samuel C. (alias) Chalfield, Samuel C.**
WIDOW: Chatfield, Charlotte 1864 Nov 29 / A-74,004 C-(blank)

024 **Chirch, Albert**
WIDOW: Chirch, Julia 1879 April 4 / A-243,472 C-(blank)

025 **Collins, Jacob**
INVALID: 1865 July 10 / A-76,663 C-(blank)

028 **Corbin, George**
INVALID: 1888 Mch [March] 8 / A-643,724 C-455,467 / Mich
WIDOW: Pashnee, Nancy 1928 July 19 / A-1,619,363 Cert. A-7-29-32 / Mich

029 **Crane, Amos**
INVALID: 1886 June 21 / A-613,733 C-879,592 / Mich
WIDOW: Crane, Charlotte 1891 Aug 5 / A-521,987 C-357,756 / Mich
MINOR: Owen, William Gdn 1893 Mar 7 / A-571,551 C-399,158 / Mich

APPENDIX A – LIST OF PENSION FILES

030 **Labosquam, Jona**
FATHER: Labosquam, John 1886 June 18 / A-341,004 C-(blank) / Mich

031 **David, John**
FATHER: Sha-wah-ne-ke-shing Oct 4, 1888 / A-381,855 C-(blank) / Mich

032 **Deland, James S.**
INVALID: 1865 Aug 21 / A-83,555 C-56,148
Note on card says: Died Oct 29, 1925, in Detroit, Mich.

033 **Driggs, William J.**
INVALID: 1890 Nov 12 / A-959,557 C-811,039 / Mich

034 **Dutton, Luther**
INVALID: 1870 Mch [March] 22 / A-154,054 C-(blank)
WIDOW: Dutton, Charlotte 1871 Jany 19 / A-193,145 C-952,031
Note on card says: Died June 1870.

035 **Etarwegeshig, John**
MOTHER1: Pohketogoquay 1869 Mar 29 / A-173,481 C-178,297
MOTHER2: Etarwegeshig, Rose 1886 Mch [March] 31 / A-338,941 C-(blank) / Mich
Note: Rose said she was John's mother and filed a claim. She did not receive a pension.

036 **Genereau, Louis**
INVALID: 1866 Feb 20 / A-103,131 C-72,070
WIDOW: Genereau, Louisa 1873 Mch [March] 29 / A-208,852 C-(blank)
MINOR: Genereau, Levi 1899 Oct 9 / A-706,520 C-(blank)

037 **George, David**
MOTHER: George, Nancy Nagatol [= Nancy Nagatoe] Oct 6, 1864 / A-67,714 C-59,838

038 **Gibson, Joseph (alias) Na-baw-na-ya-sang, Joseph**
MOTHER: Gibson, Lilla 1865 Mch [March] 25 / A-87,413 C-84,868

039 **Goeng, Samuel**
MOTHER: Tondoquay, Martha 1871 Jan 12 / A-192,965 C-(blank)

040 **Graveraet, Gerrett A.**
MOTHER: Graveraet, Sophia 1877 Mar 12 / A-230,334 C-177,583

041 **Graveraet, Henry G.**
WIDOW: Graveraet, Sophia 1865 Jan 9 / A-78,414 C-61,869

042 **Greensky, Benjamin**
WIDOW: Greensky, Emma 1864 Oct 8 / A-68,228 C-40,449

043 **Greensky, Jacob**
INVALID: 1880 July 2 / A-407,565 C-236,586 / Mich
WIDOW: Greensky, Esther 1888 June 7 / A-374,134 C-268,988 / Mich
MINOR: Block, Antoine (Gdn) 1899 Nov 14 / A-708,513 C-(blank) / Mich
Note on card says: Died June 25, 1888, Hayes Twp Charlevoix Co.

045 **Hall, Cornelius**
INVALID: 1883 Feb 6 [or Mar 6] / A-474,772 C-428,262 / Mich
Note on card says: Died 8-5-17 at Missaukee, Mich.

APPENDIX A – LIST OF PENSION FILES

046 **Hamlin, James H.**
WIDOW: Hamlin, Rosalie 1865 June 17 / A-98,298 C-76,732
MINOR: Chambers, Thomas (Guardian) 1871 Feb 6 / A-193,717 C-155,204

047 **Hannin, Joseph**
INVALID: 1890 Sept 2 / A-928,331 C-905,849 / Canada
Note on card says: Died 1-7-16 Melbourne, Canada. Attorney M. B. Stevens.

050 **Isaac, William**
INVALID: 1881 Apr 11 / A-419,031 C-348,099 / Mich

051 **Jacko, John**
INVALID: 1888 Sept 29 / A-673,904 C-896,246 / Mich
WIDOW: Jacko, Susan 1907 Jul 5 / A-872,707 C-739,295 / Mich

052 **Nai-tak-me-no-ting, Jacko (alias) Penaiswanquot, Louis (alias) Natawinoding, Louis**
INVALID: 1870 July 25 / A-158,777 C-(blank)
MOTHER: Penaiswanquot, Mary A. 1896 Dec 23 / A-645,502 C-(blank)
Note: Mary A. Penaiswonquot is the widow of Co. K soldier Jacko Penaiswonquot.

053 **Jackson, Andrew**
INVALID: 1891 Aug 11 / A-1,047,134 C-787,903 / Canada
WIDOW: Jackson, Annie S. 1894 Jan 8 / A-589,010 C-400,429 / Mich

056 **John, David**
INVALID: 1888 Sept 8 / A-671,594 C-(blank) / Mich
WIDOW: John, Mary 1894 Dec 7 / A-605,687 C-(blank) / Mich

059 **Ka-ba-ya-ce-ga, George**
INVALID: 1879 May 10 / A-285,396 C-437,815
WIDOW: Ka-ba-ya-ce-ga, Mary 1889 Dec 21 / A-411,141 C-285,353 / Mich

061 **Ka-ka-kee, Joseph (alias) Me-she-ka-kack**
INVALID: 1889 Nov 5 / A-737,101 C-619,312 / Mich

062 **Kagwaich, Samuel**
FATHER: Kagwaich, Michael 1869 Jan 29 / A-171,075 C-(blank) / Mich
MOTHER: Kagwaich, Sophia 1879 Nov 8 / A-254,020 C-369,429 / Mich
Note on card says: Died Aug 3, 1864 at Harrisburg, Pa.

063 **Kechittigo, Thomas**
INVALID: 1866 Jan 30 / A-101,234 C-264,702
WIDOW1: Kechittigo, Arton 1864 June 20 / A-54,783 C-(blank)
WIDOW2: Kechittigo, Mary 1918 Apr 2 / A-1,118,084 C-889,115 / Mich

064 **Kesses, John**
MINOR: Shipman, Eliza A. 1889 Mar 21 / A-391,344 C-(blank)
Note on card says: Died Oct 20, 1867 at Wayland, Mich. [Error, he died July 15, 1867]

066 **Ketchebatis, Amable (alias) Kechebatis, Amable (alias) Thibatis, Amable**
WIDOW: Ketchebatis, Lizette 1865 Oct 5 / A-111,210 C-128,866
MINOR: Ketchebatis, Margaret 1890 Sept 11 / A-330,708 C-(blank) / Mich

 All Rights Reserved by Chris Czopek

APPENDIX A – LIST OF PENSION FILES

068 **Kinewahwanipi, John (alias) Kenkagosa, John (alias) Ke-new-nuh-ne-ba, John (alias) Penegwah, John**
MOTHER: Kah Che, Margaret 1890 Jun 16 / A-426,267 C-(blank) / Mich
MOTHER: Banaqua, Margarette 1899 Apr 6 / A-695,639 C-498,877 / Mich
FATHER: Kah Che, John 1911 Mar 1, A-959,919 C-(blank) / Mich
SISTER: Keway, Eliza 1881 Mar 19 / A-281,385 C-(blank)

069 **Lamourandere, Thaddeus (alias) Lamorander, Taddy**
MOTHER: Lamourandere, Mary Ann
1886 Mar 29 / A-352,584 C-(blank) / Mich
FATHER: Lamorander, Aken [or Akan] 1890 Oct 16 / A-481,509 C-339,577 / Mich

072 **Marks, Louis**
FATHER: Marks, Enos 1880 Aug 18 / A-277,355 C-(blank)

074 **Mash-Kaw, James**
WIDOW: Mash Kaw, Chigahmegua 1864 July 5 / A-56,218 C-32,329

075 **Mash-Kah, John**
FATHER: Mash-Kah 1869 May 13 / A-175,170 C-196,008
MOTHER: Mash-Kah-Mak 1872 Mar 7 / A-201,624 C-163,762

076 **Miller, Thomas**
FATHER: Kahgel, H. [=Henry Kah-gee] 1870 June 14 / A-187,836 C-(blank)

078 **Miscogeom, Lewis (alias) Miscogeon, Lewis**
WIDOW: Miscogeom, Susan 1889 Jan 15 / A-387,091 C-391,056 / Mich

079 **Mesenasaw, William**
WIDOW: Mesenasaw, Mary 1865 Mch [March] 25 / A-87,267 C-66,237

080 **Mixonanby, Thomas (alias) Mixonauby, Thomas**
WIDOW: Mixonanby, Tispah 1871 May 9 / A-196,012 C-158,183

081 **Moogargoe, George W.**
INVALID: 1915 Oct 23 / A-1,421,484 C-1,177,292 / Mich
Remarks: C 2,455,410
Note on card says: Died May 16, 1923 Soldiers Home.

082 **Mwakeewenan, Daniel**
WIDOW: Mwakeewenan, Catherine 1864 Dec 3 / A-74,331 C-47,176
MINOR: Rood, Ann [Ann Rodd] (Gdn) Feb 11, 1868 / A-157,537 C-125,489

083 **Nar-quo-quot, Joseph**
WIDOW: Nar-quo-quot, Eliza 1874 April 6 / A-214,752 C-(blank)

084 **Nar-we-ge-she-quaby, Jackson (alias) Nar-ue-ge-she-gwa-by, Jackson**
MOTHER: Nar-we-ge-she-gwaby, Kaw-gay (or Kew-gey)
1892 Oct 28 / A-562,857 C-387,630 / Mich

086 **Nelson, Thomas** Units served in: K 1 Mich SS and A 14 V.R.C.
INVALID: 1866 Apr 10 / A-106,301 C-137,579
MINOR: Grewett, Peter Jr. (Gdn) 1882 Jan 30 / A-289,420 C-(blank)

 All Rights Reserved by Chris Czopek

APPENDIX A – LIST OF PENSION FILES

088 **Neveaux, William**
INVALID: 1899 Sept 5 / A-1,234,919 C-1,148,992 / Wisc
WIDOW: Neveaux, Elizabeth 1914 Nov 13 / A-1,036,907 C-788,633 / Wisc
Remarks: XC 2,688,744
Note on card says: Died Oct 16, 1914 at LaPointe, Wisc.

089 **Newton, William**
INVALID: 1890 Mch [March] 17 / A-762,351 C-1,143,892 / Mich
WIDOW: Newton, Angeline 1922 Mch [March] 5 / A-1,184,599 C-956,009 / Mich
Note on card says: Died Jan 23, 1922 at Williamsburg, Mich.

090 **Nichols, Lemuel R.** Units served in: H 7 Mich Inf , C 1 Mich SS, K 1 Mich SS
INVALID: 1877 July 23 / A-239,541 C-392,713
WIDOW: Nichols, Sarah M. 1893 Feb 6 / A-569,739 C-383,793 / Mich

091 **Noyes, Kirke W.** He served in Companies B, D and K in 1 Mich SS
INVALID: 1875 Feb 4 / A-200,670 C-327,446
Note on card says: Died August 8, 1920, South Haven, Mich.

092 **Oh-bow-ake-mo, James (alias) Ohbowakemo, James**
INVALID: 1887 Sept 30 / A-624,477 C-931,034
WIDOW: Ohbowakemo, Lucy 1896 Jan 13 / A-626,932 C-443,564
MINOR: Yuncker, Jacob (Gdn) 1899 Sept 7 / A-662,292 C-461,296

093 **Otashquabono, Leon**
INVALID: 1885 Nov 9 / A-553,786 C-363,488
WIDOW: Otashquabono, Mary A. 1902 May 5 / A-762,243 C-542,020 / Mich

094 **Otto, Marcus**
INVALID: 1865 May 11 / A-68,140 C-47,164

098 **Penaiswanquot, Jake (alias) Pasanoquoit, Jaco**
WIDOW: Penaiswanquot, Mary A. 1865 Feb 9 / A-82,393 C-263,983

100 **Pesherbay, Albert**
INVALID: 1890 Aug 12 / A-917,686 C-706,289 / Mich
WIDOW: Pesherbay, Theresa 1904 Nov 9 / A-816,581 C-616,707 / Mich

101 **Porcellay, Charles**
INVALID: 1865 Aug 9 / A-81,818 C-(blank)

103 **Quoboway, James**
INVALID: 1865 Sept 21 / A-89,918 C-75,383

105 **Sargonquatto, Aaron (alias) Pequongay, Aaron**
INVALID: 1890 Sept 20 / A-914,740 C-877,278 / Mich
WIDOW: Pequongay, Susan 1916 July 10 / A-1,069,020 C-824,040 / Mich
Remarks: XC 2,692,592. Note on card says: Died June 4, 1916 Omena.

106 **Singoby, Simon**
INVALID: 1890 Oct 20 / A-916,812 C-687,755 / Mich
WIDOW: Singoby, Mary 1903 Oct 26 / A-793,645 C-594,494 / Mich

107 **Kobanquot, Shash (alias) Kobonquot, Shash**
WIDOW: Kobanquot, Mary Shash 1864 June 20 / A-54,722 C-(blank)
[Indian Bureau Correspondence: Saline Sash-ke-bon-quet 1865 July 13 / A-101,734]

APPENDIX A – LIST OF PENSION FILES

108 **Saw-be-quom, Adam**
MOTHER: Saw-be-quom, Nah-ta-ma-ma-cy 1865 April 24 / A-90,526 C-140,550

109 **Scott, Antoine (alias) Le Croix, Antoine**
MOTHER: We-yaw-be-mind, Eliz 1890 Feb 21 / A-415,546 C-(blank) / Mich

111 **Shabena, Charles**
INVALID: 1866 Sept 6 / A-115,262 C-(blank)
WIDOW: Shabena, Mary A. 1867 Feb 19 / A-142,530 C-125,501

112 **Shanahan, Winfield S.** Units served in: K 1 Mich SS, E 1 Mich SS
INVALID: 1886 Oct 4 / A-586,519 C-(blank)
WIDOW: Shanahan, Carrie 1904 Nov 5 / A-816,433 C-754,923 / Mass

114 **Shaw-an-ese, Joseph**
WIDOW: Shaw-an-ese, Ellen 1879 Apr 4 / A-243,524 C-233,307

115 **Shawonosang, Joseph (alias) Shanernuskin, Joseph**
INVALID: 1864 Dec 27 / A-57,490 C-72,663

116 **Sheyoy, John (alias) Shegog, John**
INVALID: 1865 June 9 / A-71,336 C-59,537

117 **Shomin, John B.**
INVALID: 1892 July 31 / A-1,110,027 C-978,571 / Mich
WIDOW: Shomin, Mary M. 1913 Oct 10 / A-1,015,500 C-967,503 / Mich
Note on card says: Died Sept 16, 1913, Mackinac Is. Mich.

118 **Shomin, John**
INVALID: Dec 24, 1875 / A-211,931 C-145,119
WIDOW: Shomin, Hiacinthe Nov 4, 1895 / A-622,989 C-438,926 / Mich
MINOR: Clark, William J. (Gdn) July 5, 1899 / A-701,466 C-503,423 / Mich

119 **Shornin, Louis**
INVALID: 1880 Feb 28 / A-350,066 C-363,064
WIDOW: Shornin, Rosa 1893 May 20 / A-577,576 C-482,541
MINOR: Soyga or Sogga, Lars R. (Gdn) 1896 July 18 / A-637,596 C-504,408

120 **Smith, Thomas**
INVALID: 1866 Feb 13 / A-102,480 C-871,778
WIDOW: Smith, Nancy 1909 Apr 30 / A-919,009 C-682,003 / Mich

121 **South, Peter**
MOTHER: South, Lucy 1865 Sept 25 / A-110,435 C-127,140

122 **Stoneman, George**
INVALID: 1867 Mar 25 / A-122,832 C-110,637
WIDOW: Elliott, Mary 1921 June 4 / A-1,175,485 C-925,523 / Mich
Note on card says: Died 1887 Whitehall, Mich.

123 **Sutton, Freeman**
INVALID: 1865 Sept 27 / A-90,655 C-212,234
WIDOW: Sutton, Pemosawa 1891 May 12 / A-515,111 C-388,484 / Mich

124 **Tabasasch, Francis**
INVALID: 1882 Sept 22 / A-460,704 C-731,783

 All Rights Reserved by Chris Czopek

APPENDIX A – LIST OF PENSION FILES

125 **Tabyant, Antoine**
INVALID: 1875 June 5 / A-204,769 C-139,494
WIDOW: Tabyant, Mary A 1894 July 6 / A-598,210 C-407,371 / Mich

126 **Ta-zhe-de-win, Joseph**
INVALID: 1876 Jan 14 / A-212,999 C-142,516
Note on card says: Died Jan 3, 1923, Walkerville, Mich.

127 **Thomas, Moses**
INVALID: 1892 Aug 19 / A-1,133,166 C-922,278 / Canada
WIDOW: Thomas, Julia 1933 Nov 29 / A-1,736,502 C-(blank)
Remarks: X.C. 894915
Note on card says: Died Aug 25, 1933 at Kettle Point, Canada.

128 **Valintine, Robert**
WIDOW: Valintine, Mary 1903 Jan 9 / A-776,381 C-(blank) / Wisc

129 **Wau-bau-no, Thomas**
WIDOW: Wau-bau-no, Jane 1866 Oct 20 / A-135,921 C-(blank)

130 **Wabesis, Charles**
INVALID: 1891 Sept 14 / A-1,056,455 C-(blank) / Mich
WIDOW: Wabesis, Kutto (=Gertrude Wabsis) 1892 April 7 / A-547,165 C-(blank) / Mich

131 **Wabesis, John**
MINOR: Moh-tway-ne-ke-noh (Gdn) 1872 April 30 / A-203,469 C-167,422
Note: Moh-tway-ne-ke-noh is the name of his wife, who remarried in 1867, and is listed as Guardian of a minor child.

132 **Wakazoo, Joseph** Units served in: H 16 Mich Inf, K 1 Mich SS
INVALID: 1879 May 16 / A-288,430 C-358,213
WIDOW: Wakazoo, Necette 1910 Aug 25 / A-947,987 C-736,654 / Minn
Note on card says: Died July 18, 1910.

133 **Wassagezie, Henry**
INVALID: 1890 Oct 20 / A-998,198 C-1,080,261 / Wisc
WIDOW: Condecon, Mary 1922 Oct 19 / A-1,196,762 C-(blank) / Mich
Remarks: C 2,519,374
Note on card says: Attorney James Tanner

135 **Waubenoo, John (alias) Wanbenoo, John**
INVALID: 1895 April 1 / A-1,166,185 C-948,468 / Mich

137 **Waggishegomy, William**
INVALID: 1886 [or 1887] Feb 7 / A-596,872 C-395,277
WIDOW: Waggishegomy, Julia 1897 April 29 / A-653,176 C-454,083 / Mich

138 **Wells, Peter**
INVALID: 1888 Nov 16 / A-679,002 C-(blank) / Mich

139 **Wezoo, Thomas**
INVALID: 1884 Sept 30 / A-523,561 C-293,687 / Mich

140 **Wesley, John**
INVALID: 1889 Aug 21 / A-723,953 C-735,445 / Mich
WIDOW: Wesley, Mary 1908 Nov 25 / A-908,553 C-739,536 / Mich

 All Rights Reserved by Chris Czopek

APPENDIX A – LIST OF PENSION FILES

141 **Whiteface, Charley (alias) Now-quo-um**
FATHER: Wob-bun-day-quay-se (alias) Web-bau-day-quay-se
1874 Dec 4 / A-218,729 C-(blank)

145 **Wolf, Payson**
INVALID: 1866 May 30 / A-109,071 C-78,903

146 **Kedgnal, John (alias) Kedgenal, John** [Company] B 1 Mich SS
INVALID: 1880 Feb 28 / A-346,024 C-499,396
Note: This soldier was in Company B, not Company K.
Note on card says: Died Apr 2, 1926, Petoskey Mich.

147 **Duvernay, William** [Company] B 1 Mich SS
INVALID: 1890 July 10 / A-797,794 C-(blank) / Wisc
Note: This soldier was in Company B, not Company K.

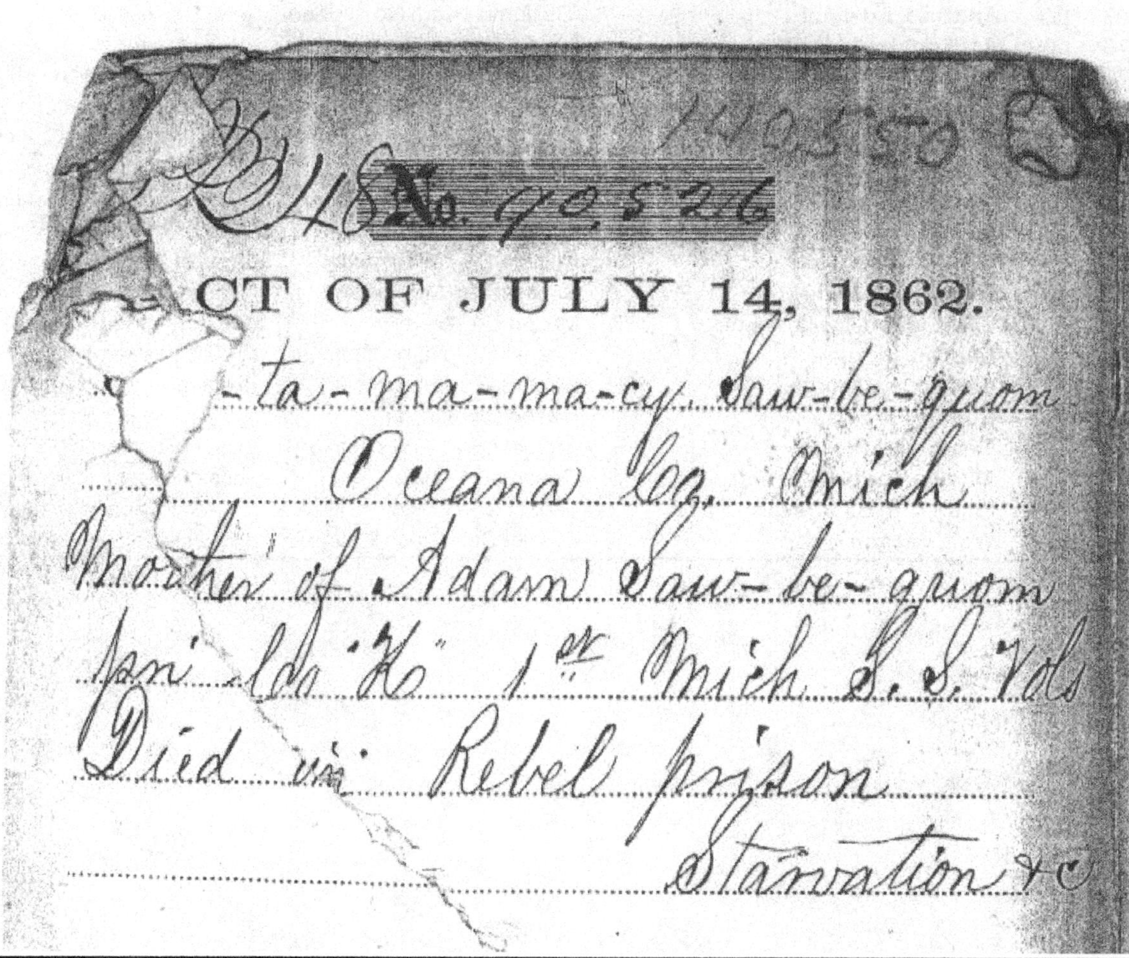
140550
No. 90526
ACT OF JULY 14, 1862.
-ta-ma-ma-cy. Saw-be-quom
Oceana Co. Mich
Mother of Adam Saw-be-quom
Pri Co "K" 1st Mich S. S. Vols
Died in Rebel prison
Starvation &c

This is part of the cover-jacket for a Civil War pension file. Adam Sawbequom died while a POW. His mother, named Na-ta-ma-ma-cy, received a pension from the Government.

APPENDIX B

Veteran Census Records

1883 Pension Roll

List of Company K soldiers and families collecting pensions as of Jan. 1, 1883.
Source: Book, titled List of Pensioners on the Roll January 1, 1883.
Note: gsw = gun shot wound.

VOL. 4	CL #	NAME on the 1883 Pension Roll	COUNTY - TOWN	REMARKS
Pg.361	001	**Agaho, Tah-gutch-quay**	Oceana Co - Hart	Widow [of Agahgo]
Pg.007	005	**Andress, Edwin V**	(California) Kern Co - Weldon	gsw r.hand & r.foot
Pg.279	009	**Ashkonois, Joseph**	Emmet Co - Cross Village	gsw left arm
Pg.279	013	**Awon, Therressa**	Emmet Co - Cross Village	Widow [of Awanakwad]
Pg.361	035	**E-tar-we-geeshig, Tah-te-go-quah**	Oceana Co - Hart	Mother [of Etarwegeshig]
Pg.279	040	**Groouat, Sophia**	Emmet Co - HarborSprings	Mother [of Garrett Graveraet]
Pg.279	066	**Ketchavatis, Elizabeth**	Emmet Co - Cross Village	Widow [of Ketchebatis]
Pg.279	074	**Mash Kay, Chic-ah-mil-qua**	Emmet Co - Cross Village	Widow [of James Mashkaw]
Pg.340	080	**Mixonanby, Tispah**	Mason Co - Custer	Widow [of Mixonauby]
Pg.308	094	**Otto, Marcus**	Isabella Co - Mt Pleasant	Loss of right arm above elbow
Pg.308	103	**Quoboway, James**	Isabella Co - Mt Pleasant	Wounded left hip
Pg.341	108	**Sarrbequon, Na-ta-ma-may**	Mason Co - Marble	Mother [of Sawbequom]
Pg.279	118	**Shomin, John**	Emmet Co - HarborSprings	gsw left forearm
Pg.360	122	**Stoneman, George**	Oceana Co - Hart	Pension granted April 1871
Pg.340	123	**Sutton, Freeman**	Mason Co - Custer	gsw left thigh
Pg.279	125	**Tab Zout, Antoine**	Emmet Co - HarborSprings	gsw right shoulder
Pg.361	126	**Tal-ghe-de-roin, Joseph**	Oceana Co - Hart	loss of little left finger
Pg.340	075	**Wash kah, Bavrg-man**	Mason Co - Custer	Father [of John Mashkaw]
Pg.279	145	**Wolf, Payson**	Emmet Co - Cross Village	Loss use left hand and arm

APPENDIX B

Veteran Census Records

1888 Veteran Census

List of Company K soldiers on the 1888 Veteran Census. Source is Reel # 684 in the RG 59-14 collection at the State Archives in Lansing.

Reel 684	CL #	NAME on 1883 Pension Roll	COUNTY - TOWN
Pg. 391	011	**Ashkebugnekay, Amos**	[Oceana] - Weldon Creek
Pg. 391	008	**Awoneghezhick, James**	[Charlevoix] - Charlevoix
Pg. 391	063	**Chettigo, Thomas K**	[Bay] - Pinconning
Note: This is Thomas Kechittigo.			
Pg. 391	094	**Otto, Marcus**	[Isabella] - Mt. Pleasant
Pg. 391	053	**Scholes [Jackson], Edward**	[Wayne] - Detroit
Note: This is Edward Andrew Jackson.			
Pg. 391	119	**Shonin, Lewis**	[Emmet] - Cross Village
Note: This is Louis Shomin.			
Pg. 391	118	**Shonin, John B**	Emmet] - Harbor Springs
Note: aka John O. Shomin, living in Harbor Springs.			
Pg. 391	120	**Smith, Thomas**	[Isabella] - Calkinsville
Note: Calkinsville was renamed Rosebush.			
Pg. 391	105	**Sargon-quatt, Aaron**	[Leelanau] - Gills Pier
Note: Gills Pier is on west shore of Leelanau peninsula near Onominisee.			
Pg. 391	105	**Schyahuahquet, Aaron**	[Leelanau] - Gills Pier
Note: Gills Pier is on west shore of Leelanau peninsula near Onominisee.			
Pg. 391	117	**Shomin, John B**	[Mackinac] - Mackinac Island
Pg. 392	009	**Tabash, Joseph**	[Emmet] - Cross Village
Note: This is Joseph Ashkanak of Cross Village.			
Pg. 392	125	**Tobiah, Antoine**	[Emmet] - Harbor Springs
Pg. 392	124	**Tabash, Francis**	[Emmet] - Cross Village
Pg. 392	126	**Ta-zhe-de-win, Joseph**	[Oceana] - Hart
Pg. 392	135	**Waw-be-noo, John**	[Isabella] - Van Decar
Pg. 392	137	**Wegesaquina, William**	[Saginaw] - Taymouth
Note: This is William Waygeshegoing.			
Pg. 392	137	**Way-ge-she-going, William**	[Saginaw] - Taymouth

 All Rights Reserved by Chris Czopek

APPENDIX B

Veteran Census Records

1890 Veteran Census

List of Company K soldiers on the 1890 Veteran Census. This a well-known source available at all large libraries in Michigan.

Census Numbers	CL #	NAME on 1890 Census	COUNTY - TOWN
SE_4, ED_227	001	**Agahgo, Charles**	**Oceana - Elbridge**
Comment: He's dead, census entry is for his widow Lucy Pequechquay.			
SE_5, ED_322	004	**Anderson, Peter**	**Mason - Ludington**
Census form comment: Discharge papers lost.			
Comment: This is CL-004 Peter Anderson, who died in 1928 in Elk Rapids, Michigan.			
SE_1, ED_079	005	**Andress, Edwin V**	**Jackson - Jackson**
Comment: He's dead, census entry is for his widow Emily E. Andress.			
SE_5, ED_061	008	**Arwonogezia, James**	**Charlevoix - Charlevoix**
Census form comment: This man was away & his wife could not talk English.			
SE_5, ED_093	009	**Ashkanak, Joseph**	**Emmet - Cross Village**
Census form comment: Wounded in arm.			
SE_5, ED_311	011	**Ashkebugnekay, Amos**	**Mason - Custer**
Census form comment: Scurvy from imprisonment in Andersonville for 9 months.			
SE_2, ED_071	093	**Aubwakoum, James**	**Lapeer - Oregon twp.**
SE_5, ED_100	013	**Awanakwad, Betois**	**Emmet - Readmond**
Census form comment: Killed [in war] No papers to show enlistment.			
Comment: He's dead, census entry is for his widow Teresa Awanakwed.			
SE_6, ED_004	015	**Bird, John (Benasis?)**	**Baraga - S_County**
Census form comment: PO: "Methodist Mission L'Anse". Sharpshooters 9 Corps.			
Comment: Who is John Bird? There was a Company K soldier named John Benasis, and the anishinabe word for bird is "Benas". I say that's too close for coincidence.			
SE_6, ED_004	029	**Crane, Amos**	**Baraga - S_County**
Census form comment: PO: "Methodist Mission L'Anse". Sharpshooters 9 Corps.			
For location it says "All of Baraga Co. South of 5th Correction line".			
SE_1, ED_090	032	**DeLand, James S.**	**Jackson - Napoleon**
Census form comment: Lost use of left arm.			
SE_x, ED_134	148	**Duvernay, William**	**Milwaukee - Nat.SoldiersHome**
Census form comment: PO: Soldiers Home, Milwaukee Wisc. Chronic Diarrhea from during Siege of Petersburg June '64.			
Comment: Official name is North West Branch of the National Soldier'sl Home, Milwaukee, Wisc.			

 All Rights Reserved by Chris Czopek

APPENDIX B

Veteran Census Records

1890 Veteran Census

Census Numbers	CL #	NAME on 1890 Census	COUNTY - TOWN
SE_5, ED_089	036	**Genereau Jr, Louis**	**Emmet - Petoskey**
Census form comment: Shot through leg. Pneumonia.			
Comment: He's dead, census entry is for his widow Louisa M. Keway (by 1890 she had remarried).			
SE_2, ED_159	137	**Gisecomey, William**	**Saginaw - Taymouth**
Comment: This is CL-138 William Waygeshegoing.			
SE_5, ED_064	043	**Greensky, Jacob**	**Charlevoix - Hayes twp.**
Comment: He's dead, census entry is for his widow Esther Greensky.			
SE_5, ED_146	045	**Hall, Cornelious W**	**Missaukee - West Branch**
SE_5, ED_270	050	**Isaac, William**	**Cheboygan - Wolverine**
Census form comment: Discharge lost, and date forgotten.			
SE_5, ED_062	051	**Jacko, John**	**Charlevoix - Evangeline**
Census form comment: Disease of lungs. Certificate of discharge in hands of board of state auditors to adjust bounty.			
SE_5, ED_319	056	**John, David**	**Mason - Sherman**
Census form comment: Papers in hands of atty [attorney], cannot get dates.			
Comment: He's dead, census entry is for his widow Mary Kawis.			
SE_5, ED_238	063	**Kechittigo, Thomas**	**Bay - Pinconning**
SE_5, ED_093	066	**Ketchebatis, Amable**	**Emmet - Cross Village**
Census form comment: Died at Andersonville. Cannot find out facts as to dates of enlistment or discharge.			
Comment: He's dead, census entry is for his widow Elizabet Ketchabaptist. She was told he died at Andersonville, but there are no records proving this.			
SE_5, ED_064	078	**Miscoguom, Louis**	**Charlevoix - Hayes twp.**
Comment: He's dead, census entry is for his widow Susan Kewagomkun.			
SE_5, ED_311	080	**Mixonauby, Thomas**	**Mason - Custer**
Comment: He's dead, census entry is for his widow Tishpah Max-no-bey.			
SE_2, ED_155	088	**Neveaux, William**	**(Wisc) Ashland - Bayfield**
SE_5, ED_141	089	**Newton, William**	**Leelanau - Leland**
SE_3, ED_099	090	**Nichols, Lemuel R**	**Shiawassee - Owosso**
Census form comment: Chronic diarrhea & piles.			
SE_4, ED_291	091	**Noyes, Kirke W**	**VanBuren - PawPaw**
SE_5, ED_063	093	**Otoshquabono, Leon**	**Charlevoix - Eveline**
SE_3, ED_028	094	**Otto, Marcus**	**Isabella - MtPleasant**
Census form comment: Loss of right arm.			
SE_5, ED_058	100	**Pesherbay, Albert**	**Charlevoix - BearLake**

Research by Chris Czopek © 2015 All Rights Reserved by Chris Czopek

APPENDIX B

Veteran Census Records

1890 Veteran Census

Census Numbers	CL #	NAME on 1890 Census	COUNTY - TOWN
SE_2, ED_157	059	**Sacketup, Gorge**	**Saginaw - StCharles**
Census form comment: Can't find out what the disability was - he drew a pension.			
Comment: He's dead, this is for his widow, Mary Sacketup. She was married to CL-059 George Kabayacega, aka George Sacketup.			
SE_5, ED_140	105	**Sargonquatto, Aaron**	**Leelanau - Leelanau**
Comment: This man is also known as Aron Pequongay. The two men are the same person.			
SE_2, ED_157	114	**Shaw-an-ese, Joseph**	**Saginaw - StCharles**
Comment: He's dead, this is for his widow, Ellen Shan-am-niss.			
SE_5, ED_095	118	**Shomin, John**	**Emmet - Friendship**
Census form comment: Gunshot wound left forearm.			
SE_6, ED_049	117	**Shomin, John B.**	**Mackinac - Mackinac Island**
SE_5, ED_093	119	**Shomin, Louis**	**Emmet - Cross Village**
Census form comment: Loss of right eyesight.			
SE_3, ED_021	120	**Smith, Thomas**	**Isabella - Isabella**
Census form comment: Blind from inflamation caused by war.			
SE_5, ED_093	124	**Tabasasch, Francis**	**Emmet - Cross Village**
SE_5, ED_095	125	**Tabiah, Antoine**	**Emmet - Friendship**
Census form comment: Gunshot wound right side of neck.			
Comment: He was actually wounded in his right shoulder.			
SE_4, ED_227	126	**Ta-zhe-de-win, Joseph**	**Oceana - Stetson**
Census form comment: Wounded - finger left hand. Finger shot off at Petersburg.			
Comment: Stetson is the old name for Walkerville in Levitt twp., Oceana Co.			
SE_4, ED_227	130	**Wabesis, Charles**	**Oceana - Elbridge**
SE_3, ED_023	135	**Waw-Be-Naw, John**	**Isabella - Nottawa**
Census form comment: Long note at bottom of page says"Waw-be-naw is an Indian and is gone for the present off of the reservation peddling baskets. I could get no satisfactory information from his son, I could not even get to see his papers if he has got them. Indians are so superstitious that it is no use to try anything except with the man himself".			
SE_1, ED_024	139	**Wezoo, Thomas**	**Calhoun - Athens**
Census form comment: Gunshot wound left side.			
Comment: He was actually wounded on the left side of his face. It caused the loss of sight in his left eye.			
SE_5, ED_093	145	**Wolf, Payson**	**Emmet - Cross Village**
Census form comment: Scurvy from being imprisoned. Prisoner in Andersonville.			

APPENDIX B

Veteran Census Records

1894 Veteran Census

List of Company K soldiers on the 1894 Veteran Census. The only information recorded was Name and Place of residence. Available at the State Library and Archives in Lansing. Look for a set of books titled Census of the State of Michigan 1894, published by Robert Smith & Co., 1896. Call number HA 442 1894. Look in volume 3 for a census of Soldiers, Sailors and Marines alive in 1894.

Page	CL #	NAME on 1894 Census	COUNTY - TOWN or TWP
Pg. 067	008	**Arwonogezice, Jas.**	**Charlevoix - Charlevoix**
Pg. 086	009	**Ashkonak, Joseph**	**Emmet - Cross Village**
Pg. 036	015	**Bird, John**	**Baraga - L'Anse**
		John Bird = John Benasis	
Pg. 123	021	**Chamberlin, Amos**	**Isabella - Denver**
Pg. 077	063	**Chitago, C. K.**	**Crawford - Grayling**
		C.K. Chitago = Thomas Kechittigo	
Pg. 123	028	**Corbin, George**	**Isabella - Denver**
Pg. 127	032	**DeLand, James S.**	**Jackson - Napoleon**
Pg. 085	146	**Deverney, John**	**Emmet - Bear Creek (=Petoskey)**
		John Deverney = John Kedgnal	
Pg. 181	045	**Hall, Cornelius W**	**Missaukee - West Branch**
Pg. 070	050	**Isaac, Wm**	**Cheboygan - Nunda twp.**
Pg. 068	051	**Jackio, John**	**Charlevoix - Hayes twp.**
Pg. 173	061	**Me-she-ka-kack, Joseph**	**Mason - Custer**
		This is Joseph Kakakee	
Pg. 095	089	**Newton, William A**	**Grand Trav. - Traverse City**
Pg. 247	091	**Noyes, Kirke W**	**VanBuren - PawPaw**
Pg. 068	093	**Otosquabona, L.**	**Charlevoix - Eveline**
Pg. 124	094	**Otto, M.**	**Isabella - Union**
Pg. 085	100	**Pasharba, Albert**	**Emmet - Bear Creek (=Petoskey)**
Pg. 156	105	**Pequongay, Aaron**	**Leelanau - Leelanau twp.**
		Aaron Pequongay = Aaron Sahgahnahquato	
Pg. 206	126	**Poneshing, Joseph**	**Oceana - Elbridge**
		Joseph Poneshing = Joseph Tazhedewin	

 All Rights Reserved by Chris Czopek

APPENDIX B

Veteran Census Records

1894 Veteran Census

Page	CL #	NAME on 1894 Census	COUNTY - TOWN or TWP
Pg. 123	103	**Quabaway, James**	**Isabella - Deerfield**
Pg. 166	117	**Shoman, John B.**	**Mackinac - Mackinac Island**
Pg. 086	118	**Shomin, John O.**	**Emmet - Friendship**
Pg. 123	120	**Smith, Thomas**	**Isabella - Isabella**
Pg. 086	124	**Tabash, Francis**	**Emmet - Cross Village**
Pg. 124	135	**Wabana, John**	**Isabella - Nottawa**
Pg. 068	008	**Wasegesik, James**	**Charlevoix - Charlevoix**
		This is James Arwonogezice	
Pg. 173	140	**Wesley, John**	**Mason - Sherman**
Pg. 057	139	**Wezo, Thomas**	**Calhoun - Athens**
		This is Thomas Wesaw	
Pg. 086	145	**Wolf, Passon**	**Emmet - Cross Village**

APPENDIX C

Places & Dates

Sorted by Home Town

When a soldier enlisted in Company K, he was asked to give his home town. In the state Capitol, Lansing, each soldier who enlisted was counted as a "draft credit" for a county in Michigan. Draft Credits were noted on the Descriptive Roll. From these two sources I have made a list of Home Towns, sorted by County. Note: John Battice, Amos Crane, and Robert Valentine came from the Upper Peninsula, but there is no record of where. Since other soldiers from the U.P. came from Ontonagon County, I listed that as their home.

HOME County	HOME Town	NAME	AGE	FATE
(CANADA Kent Co)	Harwich	Jackson, Edward Andrew	27	SURVIVED
(CANADA Lambton Co)	Walpole Island	Aptargeshick, Oliver	19	DIED
(CANADA Middlesex Co)	Oneida Reserve	Hannin, Joseph	25	SURVIVED
(MARYLAND)	Hagerstown, MD	Carter, Charles	21	DIED
(WISC Ashland Co)	La Pointe	Neveaux, William	18	SURVIVED
(WISC Ashland Co)	La Pointe	Williams, Moses	30	DIED
BARRY	Assyria	Marquette, Frank	35	SURVIVED
BAY	Kawkawlin	Johns, David	19	SURVIVED
BAY	Kawkawlin	Johns, William	20	SURVIVED
BAY	Pinconning	Sashkobanquot	45	DIED
BAY	Pinconning	Thomas, Moses	18	SURVIVED
CALHOUN	Athens twp.	Mogage, George W.	23	SURVIVED
CALHOUN	Athens twp.	Wesaw, Thomas	22	SURVIVED
CHARLEVOIX	Charlevoix	Greensky, Benjamin C.	24	DIED
CHARLEVOIX	Charlevoix	Greensky, Jacob	18	SURVIVED
CHARLEVOIX	Charlevoix	Miskoguon, Louis	30	SURVIVED
CHARLEVOIX	Charlevoix	Nauquam, Thompson	23	
CHARLEVOIX	Charlevoix	Pesherbay, Albert	25	SURVIVED
CHARLEVOIX (Manitou)	Beaver Island	Shawonosang, Joseph	18	SURVIVED
CHEBOYGAN	Burt Lake	Kejikowe, Simon	23	
CHEBOYGAN	Burt Lake	Porcellay, Charles	25	SURVIVED
CHEBOYGAN	Burt Lake	Sanequaby, Simon	25	SURVIVED

 All Rights Reserved by Chris Czopek

HOME County	HOME Town	NAME	AGE	FATE
CHIPPEWA (U.P.)	Sault Ste. Marie	Hubert, Charles	28	
EMMET	Cross Village	Ashkanak, Joseph	18	SURVIVED
EMMET	Cross Village	Ketchebatis, Amable	22	DIED
EMMET	Cross Village	Misisaius, Edward	25	
EMMET	Cross Village	Shabena, Charles	23	SURVIVED
EMMET	Cross Village	Shomin, John B.	20	SURVIVED
EMMET	Cross Village	Shomin, John O.	18	SURVIVED
EMMET	Cross Village	Shomin, Louis	19	SURVIVED
EMMET	Cross Village	Tabasash, Francis	38	SURVIVED
EMMET	Harbor Springs	Awanakwad, Petros	25	DIED
EMMET	Harbor Springs	Bennett, Louis	31	SURVIVED
EMMET	Harbor Springs	Boushaw, Augustus	20	SURVIVED
EMMET	Harbor Springs	Dabasequam, Jonah	20	DIED
EMMET	Harbor Springs	Graveraet Jr., Henry G.	55	DIED
EMMET	Harbor Springs	Graveraet, Garrett A.	23	DIED
EMMET	Harbor Springs	Kaquatch, Samuel	18	DIED
EMMET	Harbor Springs	Otashquabono, Leon	23	SURVIVED
EMMET	Harbor Springs	Tabyant, Antoine	35	SURVIVED
EMMET	Petoskey	Gibson, Joseph L.	23	DIED
EMMET	Petoskey	Marks, Louis	24	SURVIVED
EMMET	Petoskey	Mwakewenah, Daniel	40	DIED
EMMET	Petoskey	South, Peter	18	DIED
EMMET	Petoskey	Wayaubemind, Michael	19	SURVIVED
EMMET	Petoskey	Wells, Peter	18	SURVIVED
HOUGHTON (U.P.)	Pewabic	Wassagezic, Henry	17	SURVIVED
IONIA	Ionia County	Ohbowakemo, James	26	SURVIVED
ISABELLA	Isabella	Andrew, John	18	SURVIVED
ISABELLA	Isabella	Ashman, Daniel	17	DIED
ISABELLA	Isabella	Burns, Peter	21	DIED
ISABELLA	Isabella	Cabecoung, William	19	SURVIVED
ISABELLA	Isabella	Chamberlain, Amos	22	SURVIVED
ISABELLA	Isabella	Chatfield, Charles	19	DIED
ISABELLA	Isabella	Chatfield, Samuel C.	26	DIED
ISABELLA	Isabella	Collins, Jacob	30	SURVIVED
ISABELLA	Isabella	Collins, John	21	SURVIVED

 All Rights Reserved by Chris Czopek

HOME County	HOME Town	NAME	AGE	FATE
ISABELLA	Isabella	Collins, William	33	
ISABELLA	Isabella	Corbin, George	23	SURVIVED
ISABELLA	Isabella	David, John	21	DIED
ISABELLA	Isabella	George, David	22	DIED
ISABELLA	Isabella	Isaacs, John	21	SURVIVED
ISABELLA	Isabella	Isaacs, William	18	SURVIVED
ISABELLA	Isabella	Narwegeshequabey, Jackson	22	DIED
ISABELLA	Isabella	Nesogot, Joseph	27	SURVIVED
ISABELLA	Isabella	Otto, Marcus	19	SURVIVED
ISABELLA	Isabella	Otto, Solomon	22	DIED
ISABELLA	Isabella	Pemassegay, Daniel	24	
ISABELLA	Isabella	Peshekee, Mark	30	SURVIVED
ISABELLA	Isabella	Quoboway, James	32	SURVIVED
ISABELLA	Isabella	Smith, Thomas	23	SURVIVED
ISABELLA	Isabella	Wabano, Thomas	24	DIED
ISABELLA	Isabella	Waubenoo, John	22	SURVIVED
ISABELLA	Isabella	Waygeshegoing, William	22	SURVIVED
KALAMAZOO	Richland	Keeses, John	30	SURVIVED
LAPEER	Lapeer	Williams, James	18	
LAPEER	Lapeer	Williams, Samuel	20	
LEELANAU	Bingham	Prestawin, Jacob	24	DIED
LEELANAU	Grand Traverse Bay	Waukazoo, Joseph	23	SURVIVED
LEELANAU	Leelanau twp.	Jacko, John	20	SURVIVED
LEELANAU	Leelanau twp.	Kinewahwanipi, John	20	DIED
LEELANAU	Leelanau twp.	Sahgahnahquato, Aaron	21	SURVIVED
LEELANAU	Northport	Allen, Charles	19	DIED
LEELANAU	Northport	Ashkebug, George	28	SURVIVED
LEELANAU	Northport	Jacko, Nattahmenoting	21	SURVIVED
LEELANAU	Northport	Mixinasaw, William	28	DIED
LEELANAU	Northport	Penaiswanquot, Jacko	40	DIED
LEELANAU	Northport	Wolf, Payson	30	SURVIVED
LEELANAU	Omena	Miller, Thomas	19	DIED

 All Rights Reserved by Chris Czopek

HOME County	HOME Town	NAME	AGE	FATE
MACKINAC (U.P.)	St. Ignace	Hamlin, James H.	43	DIED
MACKINAC (U.P.)	St. Ignace	Jeandron, Michael	21	DIED
MACKINAC (U.P.)	St. Ignace	Kewaconda, Benjamin	21	SURVIVED
MARQUETTE (U.P.)	Marquette	Gruet, Peter	21	SURVIVED
MASON	Ludington	Sutton, Freeman	32	SURVIVED
OCEANA	Oceana Co.	Wesley, John	28	SURVIVED
OCEANA	Pentwater	Hall, Cornelius W.	18	SURVIVED
OCEANA	Pentwater	Kadah, Joseph	21	DIED
OCEANA	Pentwater	Lamourandere, Thaddeus	21	DIED
OCEANA	Pentwater	Light, Josiah	18	SURVIVED
OCEANA	Pentwater	Mixonauby, Thomas	39	DIED
OCEANA	Pentwater	Newton, William	17	SURVIVED
OCEANA	Pentwater	Watson, James V.	44	SURVIVED
OCEANA	Pentwater	Agahgo, Charles	24	DIED
OCEANA	Pentwater	Arwonogezice, James	25	SURVIVED
OCEANA	Pentwater	Ashkebugnekay, Amos	32	SURVIVED
OCEANA	Pentwater	Etarwegeshig, John	26	DIED
OCEANA	Pentwater	Genereau Jr., Louis	18	SURVIVED
OCEANA	Pentwater	Going, Samuel	20	DIED
OCEANA	Pentwater	Kabaosa, Louis	27	SURVIVED
OCEANA	Pentwater	Mashkaw, James	31	DIED
OCEANA	Pentwater	Mashkaw, John	22	DIED
OCEANA	Pentwater	Narquaquot, Joseph	25	DIED
OCEANA	Pentwater	Sawbequom, Adam	28	DIED
OCEANA	Pentwater	Scott, Antoine	22	SURVIVED
OCEANA	Pentwater	Shaw, Charles	19	SURVIVED
OCEANA	Pentwater	Shegoge, John	21	SURVIVED
OCEANA	Pentwater	Stoneman, George	22	SURVIVED
OCEANA	Pentwater	Tazhedewin, Joseph	19	SURVIVED
OCEANA	Pentwater	Wabesis, Charles	25	SURVIVED
OCEANA	Pentwater	Wabesis, John	23	DIED
OCEANA	Pentwater	Whiteface, Charles	22	DIED

HOME County	HOME Town	NAME	AGE	FATE
ONTONAGON (U.P.)	?	Battice, John	19	SURVIVED
ONTONAGON (U.P.)	?	Crane, Amos	28	SURVIVED
ONTONAGON (U.P.)	?	Valentine, Robert	23	SURVIVED
ONTONAGON (U.P.)	Algonquin	Benasis, John	22	SURVIVED
ONTONAGON (U.P.)	Greenland	Seymour, Joseph	20	SURVIVED
ONTONAGON (U.P.)	Rockland	Amderling, Peter	26	SURVIVED
ONTONAGON (U.P.)	Rockland	Rubberdee, John	25	SURVIVED
OTTAWA	Chester	Kakakee, Joseph	35	SURVIVED
OTTAWA	Grand Haven	Duvernay, William	13	SURVIVED
OTTAWA	Grand Haven	Kedgnal, John	18	SURVIVED
SAGINAW	Chesaning	Shawanese, Joseph	23	SURVIVED
SAGINAW	East Saginaw	Kechittigo, Thomas	27	SURVIVED
SAGINAW	Saginaw	Jackson, William	26	SURVIVED
SAGINAW	Saginaw	Kabayacega, George	30	SURVIVED
SAGINAW	St. Charles	Church, Albert	20	SURVIVED
SAGINAW	Taymouth	Dutton, Luke	19	SURVIVED
SAGINAW	Taymouth	Nelson, Thomas	27	SURVIVED
ST.JOSEPH	Mendon	Lidger, David	28	SURVIVED

APPENDIX C

Places & Dates

Sorted by Enlistment Date & Place

Note: Buena Vista = East Saginaw. Little Traverse = Harbor Springs. Bear River = Petoskey.

ENLISTMENT DATE	ENLISTMENT PLACE	NAME (Prime Name)
11/08/61	Halls Hill, Va.	Waukazoo, Joseph
12/04/62	Grand Haven	Duvernay, William
12/10/62	Grand Haven	Kedgnal, John
05/03/63	Buena Vista	Kechittigo, Thomas
05/10/63	East Saginaw	Burns, Peter
05/11/63	Isabella	Isaacs, William
05/18/63	Isabella	Andrew, John
05/18/63	Isabella	Ashman, Daniel
05/18/63	Isabella	Cabecoung, W.
05/18/63	Isabella	Chatfield, Charles
05/18/63	Isabella	Chatfield, Samuel
05/18/63	Isabella	Collins, John
05/18/63	Isabella	Collins, William
05/18/63	Isabella	Corbin, George
05/18/63	Isabella	George, David
05/18/63	Isabella	Kargayarsang, S.
05/18/63	Isabella	Nelson, Thomas
05/18/63	Isabella	Otto, Marcus
05/18/63	Isabella	Otto, Solomon
05/18/63	Isabella	Pemassegay, Dan
05/18/63	Isabella	Quoboway, James
05/18/63	Isabella	Sashkobanquot
05/18/63	Isabella	Smith, Thomas
05/18/63	Isabella	Wabano, Thomas
05/18/63	Isabella	Waubenoo, John

ENLISTMENT DATE	ENLISTMENT PLACE	NAME (Prime Name)
05/18/63	Isabella	Waygeshegoing, William
05/19/63	East Saginaw	David, John
05/19/63	East Saginaw	Peshekee, Mark
05/19/63	Port Huron	Isaacs, John
05/29/63	Dearborn	Williams, James
05/29/63	Dearborn	Williams, Samuel
06/01/63	Bear River	Marks, Louis
06/01/63	Northport	Penaiswanquot, J.
06/02/63	East Saginaw	Dutton, Luke
06/05/63	Isabella	Nesogot, Joseph
06/08/63	Little Traverse	Awanakwad, P.
06/08/63	Little Traverse	Sanequaby, Simon
06/09/63	Little Traverse	Graveraet, Garrett
06/10/63	East Saginaw	Church, Albert
06/10/63	East Saginaw	Collins, Jacob
06/11/63	Cross Village	Shabena, Charles
06/11/63	Cross Village	Shomin, John O.
06/11/63	Cross Village	Shomin, Louis
06/11/63	Cross Village	Tabasash, Francis
06/12/63	Little Traverse	Bennett, Louis
06/12/63	Northport	Allen, Charles
06/13/63	East Saginaw	Chamberlain, A.
06/13/63	East Saginaw	Shawanese, Joseph
06/13/63	Northport	Miller, Thomas
06/14/63	Little Traverse	Tabyant, Antoine

 All Rights Reserved by Chris Czopek

ENLISTMENT DATE	ENLISTMENT PLACE	NAME (Prime Name)
06/14/63	Northport	Misisaius, Edward
06/15/63	Little Traverse	Kaquatch, Samuel
06/17/63	Bear River	South, Peter
06/17/63	Bear River	Wayaubemind, M.
06/17/63	Dearborn	Aptargeshick, O.
06/17/63	Little Traverse	Wells, Peter
06/18/63	Little Traverse	Pesherbay, Albert
06/18/63	Little Traverse	Shawonosang, J.
06/23/63	Dearborn	Andress, Edwin V.
06/24/63	Isabella	Narwegeshequabey
06/24/63	Little Traverse	Dabasequam, Jonah
06/24/63	Little Traverse	Nauquam, T.
06/28/63	Little Traverse	Boushaw, A.
07/04/63	Elbridge	Agahgo, Charles
07/04/63	Elbridge	Ashkebugnekay, Amos
07/04/63	Elbridge	Genereau Jr, Louis
07/04/63	Elbridge	Going, Samuel
07/04/63	Elbridge	Kabaosa, Louis
07/04/63	Elbridge	Lamourandere, T.
07/04/63	Elbridge	Light, Josiah
07/04/63	Elbridge	Mashkaw, James
07/04/63	Elbridge	Mashkaw, John
07/04/63	Elbridge	Narquaquot, Joseph
07/04/63	Elbridge	Newton, William
07/04/63	Elbridge	Sawbequom, Adam
07/04/63	Elbridge	Scott, Antoine
07/04/63	Elbridge	Shaw, Charles
07/04/63	Elbridge	Shegoge, John
07/04/63	Elbridge	Stoneman, George
07/04/63	Elbridge	Tazhedewin, J.
07/04/63	Elbridge	Wabesis, Charles
07/04/63	Elbridge	Wabesis, John
07/04/63	Elbridge	Watson, James V.
07/04/63	Elbridge	Whiteface, Charles
07/04/63	Marquette	Williams, Moses
07/05/63	Elbridge	Etarwegeshig, J.
07/06/63	Dearborn	Gruet, Peter
07/06/63	Little Traverse	Kejikowe, Simon
07/06/63	Little Traverse	Porcellay, Charles
07/07/63	Dearborn	Graveraet, Henry
07/08/63	La Pointe. Wisc	Neveaux, William
07/09/63	Northport	Ashkebug, Geo.
07/09/63	Northport	Mixinasaw, W.
07/14/63	Traverse	Ashkanak, Joseph
08/03/63	Northport	Wolf, Payson
08/09/63	Detroit	Hubert, Charles
08/15/63	Bear River	Mwakewenah, Dan
08/17/63	Grand Traverse	Shomin, John B.
08/22/63	Northport	Otashquabono, L.
08/24/63	Northport	Ketchebatis, A.
09/08/63	Little Traverse	Gibson, Joseph L.
09/09/63	Elbridge	Kadah, Joseph
09/10/63	Little Traverse	Greensky, Ben
09/10/63	Little Traverse	Miskoguon, Louis
09/12/63	Little Traverse	Arwonogezice, J.
02/01/64	Kalamazoo	Ohbowakemo, J.
02/01/64	Kalamazoo	Wesley, John
02/23/64	Kalamazoo	Mogage, George
02/24/64	Mackinac	Kewaconda, Ben
03/14/64	Mackinac	Jeandron, M.
03/28/64	Mackinac	Hamlin, James H
06/22/64	Petersburg VA	Carter, Charles
08/21/64	Grand Rapids	Kakakee, Joseph
08/23/64	Kalamazoo	Keeses, John

 All Rights Reserved by Chris Czopek

ENLISTMENT DATE	ENLISTMENT PLACE	NAME (Prime Name)
08/24/64	Jackson	Battice, John
08/24/64	Jackson	Crane, Amos
08/24/64	Jackson	Valentine, Robert
08/29/64	Flint	Jackson, William
08/29/64	Flint	Kabayacega, Geo.
09/02/64	Grand Rapids	Greensky, Jacob
09/03/64	Grand Rapids	Hall, Cornelius W
09/03/64	Grand Rapids	Mixonauby, Thom.
09/24/64	Greenland	Seymour, Joseph
09/24/64	Pewabic	Wassagezic, Henry
09/24/64	Van Buren twp.	Wesaw, Thomas
09/26/64	Rockland	Amderling, Peter
09/26/64	Rockland	Rubberdee, John
10/01/64	Detroit	Jackson, Edward A.
10/10/64	Algonquin	Benasis, John
10/24/64	Grand Rapids	Marquette, Frank
11/18/64	Traverse City	Prestawin, Jacob
01/27/65	Detroit	Thomas, Moses
01/27/65	LaGrange	Johns, David
01/27/65	White Pigeon	Johns, William
02/03/65	Mendon	Hannin, Joseph
02/04/65	Detroit	Lidger, David
02/10/65	Grand Rapids	Sutton, Freeman
02/14/65	Grand Rapids	Jacko, John
02/14/65	Grand Rapids	Kinewahwanipi, J.
02/14/65	Grand Rapids	Sahgahnahquato, Aaron

MISCELLANEOUS

OBITUARIES

OLD INDIAN PATRIOT DIED LAST THURSDAY

Jim "Lightsky" Served Through Rebellion In 1st Michigan Inft.

An old Indian soldier, 80 years of age Jim Wosegesic (in English "Lightsky") died Thursday of last week at his home in Bayshore, of pneumonia, leaving a wife and a stepson.

Wosegesic was a member of the 1 and served through the Civil war with great credit. He has been preaching for the last 51 years among the Indian people and was noted for his kindness and generosity to the sick and the poor.

He was buried Saturday in the Indian cemetery at Susan Lake. Funeral services were held Saturday morning at the M. E. church in this city, Rev. William Petoskey of the Indian church officiating, assisted by Rev. Quinton Walker.

Charlevoix Courier
Wed, May 23, 1917, pg5

On the third of April occurred the death of Leo Odesheweiad an Indian living near Charlevoix. He was an ex-soldier of the Civil war, having served three years with Company K of the 1st Michigan sharpshooters, and was a member of Baxter Post No. 119, Charlevoix. He was taken by a detail of his comrades to Harbor Springs for burial in the Catholic cemetery Saturday.

Petoskey Record
April 9, 1902

VETERAN DIES—A Civil War veteran, William C. Knowlton, an aged Indian resident of Yuba, died this week. Funeral was held Wednesday and burial was made in the Yuba cemetery.

Traverse City Record Eagle
January 28, 1922 Pg.3

—Mrs. Theresa Pershabay, an old and respected Indian woman, passed away Friday afternoon at her home in Bear Creek township. She was the widow of Albert Pershabay, who was an honored veteran of the Civil war. The funeral will be held Sunday at 1:30 at St. Francis church, interment in Greenwood.

from the Petoskey Evening News
Saturday, June 6, 1908, pg 3

SAR-GON-QUAT-TO

His American name was Aaron Poquongay and he was a member of Company K, 1st Michigan Sharpshooters and served in the war between the states. He was the son **of Nash-kaze, born in 1837, and was with Sherman in his march to the sea. He is buried in the township cemetery.**

from newspaper article in Traverse City Record-Eagle Oct 29, 1955, information from W.A. Craker, pioneer who knew the Indians in Omena.

 All Rights Reserved by Chris Czopek

MISCELLANEOUS
OBITUARIES

The death of Francis Tabasash at Cross Village, Friday, marks the passing of the most unique figure in the G. A. R. in that vicinity, if not in the state. Though an Indian, he fought in many of the fiercest battles of the Civil war and was a sergeant in company K, First Michigan sharpshooters. He was born at Cross Village 87 years ago and enlisted from that place.

Clipping from unknown newspaper, c. Nov. 1912

James Wau-sha-gee-sic of the Soldiers Home, Grand Rapids, arrived on Saturday and will spend several weeks in the north on a furlough, returning again on May 29. He is a local preacher and served in the First Michigan regiment of Sharp Shooters in Company K, made up entirely of Indien soldiers who went through their war records with an eye which knew its mark and never failed. Wau-sha-gee-sic today looks as though he could pick an eagle from its loft with as true an aim as his white brethren pick off mud pies at fifty feet.

Traverse City Record Eagle - April 20, 1905

Oldest Indian Dies—

Mrs. James Watsejesick, 100 years old, claimed to be this region's oldest native Indian, died at her residence in Hayes township Sunday. Mrs. Watsejesick's first husband, Sam Topkekae, is said to have been a sharp-shooter in the Civil War, leading a company of Indians he organized from this region. He was later a preacher at Greensky Hill Indian mission, where funeral services for the aged woman were held today.

Charlevoix News
February 14, 1933

INDIAN VETERAN DEAD

Was Member of Northern Michigan Indian Sharpshooters

Sunday morning, at Cross village, says the Harbor Spring's Republican, Joseph Aschaunach, a veteran of the civil was, answered his last roll call and joined his comrades in arms who in countless thousands have been enrolled in the legions of the Great Commander in the land that can be reached only through the portals of the grave.

Deceased was a member of Co. K 1st Michigan Sharpshooters, a company of Indians organized in this vicinity by Lieut. Garret A. Graveraet, brother of Mrs. Roseine Wright of Harbor Springs the young man at that time being the government school teacher at Middle Village. This was one of the two companies of Indians who served in the Union armies; and the only one enlisted from this state, the other being a company organized in the state of New York. Of the 100 members of Co. K. 42 were from Cross Village. The others went from Good Hart, Little Traverse and other points in Northern Michigan: Only one other member of the original company is left, John S. Shomin, who lives on Mackinac Island.

Charlevoix County Herald
June 13, 1914

Amos Green, an aged and respected Indian of Custer was laid to rest last week in the Custer cemetery. Amos was a civil war veteran and several members of S. D. Haight G. A. R. of this place attended the funeral. Among those were comrades J. N. Clark, N. I. Kinney, E. W. Loomis, Thos. Barnes, Rev. Arthur and N. T. Holmes.

LUDINGTON CHRONICLE
Wed. June 27, 1906 pg.5 col.3
Haight GAR Post No. 348 Scottville, Mich.

MISCELLANEOUS

THE STORY OF CHIEF MWAKEWENAH

A Michigan Indian Sharpshooter.

Those who saw the passage through Washington of Gen. Burnside's corps will remember the Indian battalion embraced in it, a body of active, robust young men. They were volunteers from the Ottawas of Michigan, and constituted the 1st Michigan sharpshooters. In the fight in the Wilderness this regiment was deployed as skirmishers, and made sad havoc in the rebel ranks, especially among their officers. One of the most noted of the Ottawas was a chief named Mo-ke-we-naw, or as better known to his neighbors in Michigan, Daniel Mo-ke-we-naw. He was very popular with the tribe, and was the principal in enlisting the band. He was very skillful in the use of his rifle, and his men say that in the first fight in the Wilderness he killed not less than thirty-two rebels, a number of them officers. Towards the close of the fight, a rebel sharpshooter selected Daniel as a mark—a very prominent one, by the bye, he standing six feet four inches in his moccasins, and proportionately large in frame. The rebel fired, and the ball took effect in Daniel's hand, and he being unable to reload his rifle, left the field. He went to Fredericksburg with the wounded, and from thence came to Washington. It was decided by the surgeons that a portion of the wounded hand must be amputated. The operation was performed, but death ensued. The body was taken in charge by Hon. John F. Driggs, of Michigan, who had it embalmed by Dr. Thomas Holmes and placed in a handsome coffin, with a plate, upon which was inscribed the name of the chief, the cause of his death, the date and name of the battle in which it was received. The body was then placed in charge of [illegible] to Bear [illegible].—*Washington Star*, 16th.

The remains of the Indian hero arrived in this city a few days since, and, as there was no means of giving timely information to his people, was temporarily buried in the City Cemetery—doubtless to be removed, at some future time, to the burial place of the Ottawas.

One of the most interesting soldiers in Company K was Chief Mwakewenah. At the age of 40 he enlisted in the Union Army and fought in the Civil War. He died from wounds received at the Battle of Spotsylvania.

Because of his status as Chief of the Bear River band (near Petoskey), Congressman Driggs arranged for his body to be sent home to his people. He was buried in Brady Hill cemetery, in downtown Saginaw.

Notice the newspaper story on the left calls him "the Indian Hero".

← *From: Saginaw Enterprise, June 30, 1864*

Garrett A. Graveraet

Native American Army Officer
Company K
1st Michigan Sharp Shooters

Died in the Civil War from wounds received in battle. Buried on Mackinac Island.

After the death of Lieutenant Graveraet, an obituary appeared in Detroit's largest newspaper, the Advertiser and Tribune. It was written by Dewitt C. Leach, the Michigan Commissioner of Indians, who was personally acquainted with Garrett and his family. Here are some of the things he said:

Lieutenant Garrett A. Graveraet was, I think, a native of Mackinac. His father was a white man; his mother an Indian woman of the Chippeway tribe.
Young Graveraet's early life was spent in the wild region around Mackinac Island. Hence his faculties for education were very limited, but so well were they improved that he became a very good scholar. He spoke the English, French and Chippeway language with care and fluency. He had a taste for painting, and, without instruction, executed some very creditable pieces.

My acquaintance with him commenced about three years ago. About that time he received an appointment as teacher in an Indian school near Little Traverse.
He faithfully performed his duties as teacher, but the confinement of the school room proved injurious to his health, and compelled him to resign.
Soon after his resignation, authority was given to raise a company of Indian volunteers for the 1st Michigan Sharp Shooters. Patriotic and fearless in the highest degree, he entered with energy and zeal upon the recruiting service, and when the regiment was formed, was commissioned Second Lieutenant of Company K.

I last saw Lieutenant Graveraet at Annapolis, Maryland, last March. He was then, although not in good health, busy with hammer and saw making repairs in camp, and hopeful and cheerful as ever.

(continued on next page)

After the regiment went to the front he was engaged in several severe battles, and upon all occasions proved himself more than worthy of the place he filled.

His father, who was a Sergeant in the same company, was killed in the terrible battle at Spotsylvania. The Lieutenant, who was noted for his affection for his parents, had no time for tears. Beside his father's corpse he fought with renewed energy, apparently regardless of his own safety, and seeking only his country's honor and welfare.

Lieut. W. J. Driggs, of the same company, says of him that "he never shrank from danger. At the battle of the Wilderness, and at Spotsylvania he remained constantly with the men, sharing their danger and encouraging them to stand their ground. He often complained of his health, which was usually feeble, and I had frequently urged him to remain at the hospital or in camp. But this he would not do. He was determined to be with his men. In the disastrous charge before Petersburg, on the 17th of June [1864], he placed himself at the head, and called on them to follow. It was there a bullet struck his left arm near the shoulder. He was taken to the hospital and the arm amputated. The next day he was taken to Washington where he died, I think, on the 1st of July. He was the most popular officer in the regiment, and his loss was deeply felt."

As a fitting close to this brief notice of one of the most worthy young men I have ever known, I will add that his last act before leaving for the seat of war, was to purchase and fit up a house and lot for his mother and sister, who were chiefly dependent on him for support … I have said his mother was an Indian woman; but she is an educated and worthy one. She has given her husband and her only son to the country. May a merciful Providence and a generous people watch over her declining years!

One thing more. Lieut. Graveraet spent the greater part of his life among a people somewhat rough and uncultivated; but he was gentlemanly and refined in his manners. He moved in a circle where intemperance was common, but he touched not the intoxicating cup. He was surrounded by profanity, but he was not profane.

After the war, the body of Lieut. Graveraet was brought back to Michigan and buried in historic St. Anne Cemetery on Mackinac Island.

On the front lawn of the State Capitol building in Lansing is a monument to the 1st Michigan Sharpshooters regiment. Graveraet's name is on that monument.

MISCELLANEOUS

ANDERSONVILLE PRISON and COMPANY K

During the Siege of Petersburg, 15 men from Company K were captured in battle on June 17, 1864, and sent to a prison camp in Georgia. The correct name for their destination was Camp Sumter, but everyone called it **Andersonville Prison**.

LIST OF MEN SENT TO ANDERSONVILLE, AND THEIR FATE

Ashkebugnekay, Amos (aka Amos Green)	Home: Elbridge	SURVIVED
Ashman, Daniel	Home: Isabella	Died while a POW
Gibson, Joseph L.	Home: Petoskey	Died – Grave 7741
Hamlin, James H.	Home: St. Ignace	Died – Grave 11260
Jeandron, Michael	Home: St. Ignace	Died – Grave 12396
Marks, Louis	Home: Petoskey	SURVIVED
Miskoguon, Louis	Home: Charlevoix	SURVIVED
Mixinasaw, William	Home: Northport	Died – Grave 11511
Nesogot, Joseph (aka Joseph Williams)	Home: Isabella	SURVIVED
Newton, William	Home: Elbridge	SURVIVED
Penaiswanquot, Jacko	Home: Northport	Died – Grave 11705 (Recalt)
Sawbequom, Adam	Home: Elbridge	Died – Grave 11508
Shomin, John B.	Home: Cross Village	SURVIVED
South, Peter	Home: Petoskey	Died – Grave 12310
Wolf, Payson	Home: Northport	SURVIVED

TRADITIONAL CEREMONY HONORS THE SEVEN GRAVES

On May 15, 2010, the seven graves of the Company K soldiers buried at Andersonville were honored with a traditional ceremony by a group of Native American veterans from Michigan. This historic event was conducted by the Anishinabe Ogitchedaw Veterans and Warriors Society, a veterans group that meets in Mt. Pleasant on the reservation of the Saginaw-Chippewa tribe.

JACKSON NARWEGESHEQUABEY

Official records state he was captured at the Crater on July 30, 1864. However, genealogist Quita V. Shier says he was captured on June 17, 1864, with the other 15 men and sent to Andersonville prison. See SHIER, Warriors in Mr. Lincoln's Army, pg 263.

AMABLE KETCHEBATIS was also a POW. Captured July 30, 1864 at "the Crater". Died February 22, 1865 in prison at Danville, VA.

MISCELLANEOUS

GRAVES OF COMPANY K SOLDIERS AT ARLINGTON NATIONAL CEMETERY

Washington, D.C.

DAVID GEORGE

Residence: ISABELLA

Died on May 26,1864, of wounds received at Spotsylvania, VA

BURIED: Section 27, Grave # 325

Directions to grave: Walk on Ord & Weitzel drive starting at the Custis Walk and go towards Lodge No.2. On your left look for a military style gravestone for Thomas Newman. When you find Newman, he is the first stone in a row. Go down this row and grave number 325 will be David George.

OLIVER APTARGESHICK

Residence: WALPOLE ISLAND, CANADA

Died of wounds July 9,1864, in a hospital in Washington DC

BURIED: Section 13, Grave # 5732

Directions to grave: Walk on Meigs Drive starting at Wilson drive and walk towards McPherson Drive. On your left look for a large, civilian style gravestone for John Joy Edson. When you find Edson, he is the first stone in a row. Walk down this row for 180 paces and look for grave number 5732.

PETER BURNS

Residence: ISABELLA (Enlisted at East Saginaw)

Died of disease September 14, 1864, in a hospital in Washington DC

BURIED: Section 13, Grave # 8141

Directions to grave: Walk on McPherson Drive starting at Miegs drive and walk towards the Section 14 circle. On your left look for a gravestone marked "Unknown Confederate soldier". This confederate stone is the first stone in a row. Walk down this row and look for grave number 8141.

Only Surviving Widows of Indian Civil War Veterans

MRS. S. A. PEQUANGAY — MRS. WM. NEWTON

TRAVERSE CITY, Jan. 5—Two widows, Mrs. William Newton of Honor, and Mrs. Susan Allen Pequangay of this city, Ottawas, are the oly surviving widows of Michigan Ottawa Indians who served in the Civil war.

Company K, first regiment of Michigan Sharpshooters, was composed entirely of Indians who were identified with 19 battles, taken prisoners before Petersburg and marched into Libby prison, later transferred to Andersonville prison, where, after untold horrors, they were exchanged as prisoners and sent north on the steamer Sultana. Later many of the feeble survivors succumbed to their prison treatment.

In the death of William Newton four years ago, taps sounded for the entire company of 101 soldiers, who served under Capt. Edwin Andrus, Maj. Riper, Col. Charles V. DeLand and Gen. Allen Burnside.

Mrs. Pequangay, widow of Aaron Pequangay, was educated in the Peter Daugherty school at Old Mission and though nearly '80, is robust and strong. She is a noble example of the old time educated Ottawa. Mrs. Newton, 80, quaint and shy, speaks no English but depends on her granddaughter, Elizabeth Wassaquom, to do her interpreting. Both draw pensions of $30 a month, which provides for all their wants. Each widow owns a modest home and is content.

Mrs. Pequangay was the widow of Aaron Sahgahnahquato. Mrs. Newton was the widow of William Newton, who died on January 27, 1922. (This news story was printed in 1926).

Note: Unknown to the writer of this story, there was one other widow still living – Mrs. Elizabeth Neveaux, resident of La Pointe, Wisconsin.

MISCELLANEOUS

MONUMENT TO THE 1st MICHIGAN SHARPSHOOTERS IN LANSING

This monument sits on the front lawn of the State Capitol building in Lansing. It was dedicated in 1915 as a memorial to the 1st Michigan Sharpshooters regiment, of which Company K was a part. The name of Lieutenant Garrett Graveraet is inscribed on one of the bronze tablets surrounding the base.

 All Rights Reserved by Chris Czopek

ABOUT THE AUTHOR

Chris Czopek

Lansing, MI

Soldier - U.S. Army, Military Intelligence

Archaeologist - Volunteer for the 1987 dig at the Roman city Caesarea (Israel)

Photographer - Michigan Senate News Office

State Worker - MSP Records Division

Historian – Member Sons of Union Veterans Curtenius Guard Camp 17, Sunfield, MI

Highlights of his career as a Civil War historian:

- **Marshall, MI : GAR Hall Research Project and Rededication in 2002**
- **Andersonville, GA : The Michigan Monument Rededication in 2004**
- **Discovered the story of Morgan Dowling, Andersonville survivor**
- **Delta Cannon Research Project (Delta township cemetery, Eaton Co.)**
- **Consultant for Documentary shown on The History Channel in 2006 titled "Indian Warriors, the Untold Story of the Civil War".**
- **Assistant Producer and lead speaker in the award-winning film documentary titled "Road To Andersonville", premiered April 2013.**
- **Advocate for the recognition of Company K, 1st Mich. Sharpshooters**

Chris grew up in Michigan during the Civil War Centennial. Stories in LIFE magazine and special television programs captured his imagination and started a lifelong interest in Civil War history. As a boy, he would watch a movie on TV – then go to the library and find out how much of the story was true. This was the beginning of his career as a Civil War historian.

While making a list of Civil War soldiers buried in Calhoun County, he discovered the grave of Thomas Wesaw on a reservation near Athens. It was marked with a standard Civil War soldier gravestone and gave "Co. K, 1 Mich SS" for the unit. Researching this soldier led him to discover the story of Company K.

Made in the USA
Monee, IL
13 September 2026

62979546R00125